Design Synthesis:

Integrated Product and Manufacturing System Design

Design Synthesis:

Integrated Product and Manufacturing System Design

Graeme Britton

Seppo Torvinen

CRC Press
Taylor & Francis Group
Boca Raton   London   New York

CRC Press is an imprint of the
Taylor & Francis Group, an **informa** business

CRC Press
Taylor & Francis Group
6000 Broken Sound Parkway NW, Suite 300
Boca Raton, FL 33487-2742

First issued in paperback 2018

© 2014 by Taylor & Francis Group, LLC
CRC Press is an imprint of Taylor & Francis Group, an Informa business

ISBN 13: 978-1-138-07374-6 (pbk)
ISBN 13: 978-1-4398-8164-4 (hbk)

Library of Congress Cataloging-in-Publication Data

Britton, Graeme Arthur.
 Design synthesis : integrated product and manufacturing system design / Graeme Arthur Britton, Seppo Torvinen.
  pages cm
 Includes bibliographical references and index.
 ISBN 978-1-4398-8164-4 (hardback)
 1. Production engineering--Textbooks. 2. Product design--Textbooks. I. Torvinen, Seppo J. II. Title. III. Title: Integrated product and manufacturing system design.

TS176.B75 2013
670--dc23   2013017993

Visit the Taylor & Francis Web site at
http://www.taylorandfrancis.com

and the CRC Press Web site at
http://www.crcpress.com

Contents

Section I Foundation

Section II   Context

Section III Synthesis—Technologies

Section IV   Synthesis—People

List of Illustrations

List of Tables

Preface

Small and medium companies employ more people and contribute a larger share to GDP than large companies. Furthermore, they tend to be innovative and adaptive, and they can be global market leaders as well. For all these reasons, small and medium companies are vitally important to national economies. It is unfortunate, therefore, that so little attention has been paid to these companies in the engineering and business literature.

The authors have worked with small and medium engineering companies over many years in several countries. We have helped them develop innovative products and processes. Innovative products require major changes in the manufacturing processes, especially assembly methods. One of the most difficult challenges faced by companies is jointly designing innovative products and the manufacturing systems to make them. In this book, we describe and explain how products and manufacturing systems can be jointly designed through architectural integration and design life cycle harmonisation. The key benefits of our book lie in its wide scope of coverage, depth of coverage, multi-disciplinary perspective and presentation of research and practices that are difficult or almost impossible to find. Readers can use our book to jointly optimise the design/re-design of new products and manufacturing systems to produce better quality products faster and cheaper.

Graeme Britton and Seppo Torvinen

Acknowledgements

The authors thank the National Technology Agency of Finland (TEKES) for funding the research projects—PlanetOne, PlanetTwo and MSDD ramp-up—which provided the research material for the book. The authors also acknowledge Tampere University of Technology and Nanyang Technological University for supporting them in their research.

We thank the many small and medium Finnish manufacturing companies that participated in the research projects. Special thanks are due to Juho Nummela, CEO of Ponsse, for his contributions to the book.

The Authors

Graeme Britton, BE (Hons), PhD, DEng, CEng, FIMechE, SnrMSME, has over 30 years' academic experience in universities in Finland, New Zealand and Singapore, and extensive knowledge of the tertiary education sectors in Australia, China, Finland, Malaysia, New Zealand and Singapore. Currently he is a vice president in Raffles Education Corporation, the largest private tertiary education provider in the Asia Pacific region. He is responsible for corporate development, coordination of the universities in the Raffles network and academic quality assurance.

Professor Britton has many years' experience conducting applied research in industry, including company-specific projects, national projects and international collaborative research projects. His research and consultancy work has been focused around helping small and medium engineering companies. He has published over 100 refereed articles and book chapters for which he was awarded the doctor of engineering (DEng).

Seppo Torvinen is a professor of production engineering at Tampere University of Technology, Finland. His areas of interest and expertise are production systems design, demand-driven supply networks, industrial information systems, sustainability and socio-technical systems design. During his academic career, Professor Torvinen has served in the following institutions abroad:

- Norges teknisk-naturvetenskaplige universitet, Trondheim, Department of Production Engineering, Norway, visiting researcher 1988–1989. Main research topics: Distributed control systems for flexible manufacturing systems, tool management systems.
- Massachusetts Institute of Technology, Production System Design Laboratory, Cambridge, visiting professor 2001–2002. Main research topics: Design of production system structure, product design for producibility.
- Nanyang Technological University, School of Mechanical and Production Engineering, Singapore, visiting professor 2002–2003. Main research topics: Design of production system structure, product design for producibility, design of productive working environments.
- National Institute of Standards and Technology (NIST), Gaithersburg, Maryland, 2010. Main research topics: Sustainable manufacturing, interaction of product and production system life cycles.

Professor Torvinen is active in international collaboration and was appointed by the Estonian Higher Education Accreditation Centre to act as chairman of the International Committee evaluating the curricula and study programs of Estonian universities in the area of mechanical engineering. He was also appointed to act as a strategic advisor in manufacturing engineering for the government of the Kingdom of Sweden during the month of May 2009. During his several sabbatical leaves, Professor Torvinen has advised and still advises several Finnish world-class companies and, among other companies, has acted as a technology director at the Formia Technology Group Ltd. and Profile Vehicles Ltd. He is also member of the board of directors in several companies. He has been working with the European Commission as evaluator, rapporteur and reviewer since 1995, starting with the ESPRIT program until the present FP7. Professor Torvinen acted as an independent observer in the monitoring of European Commission's FP7 ICT 3rd Call in 2008 and 5th Call in 2010.

Section I

Foundation

1

Introduction

The first two sections of the introduction discuss the purpose of the book and the target audience. Next, two methods for improving design learning—reflective journals and concept maps—are described and explained. The methods are intended to improve the rate and effectiveness of your learning as you work through the book chapters.

Next, there is a short introduction to global manufacturing, which provides a background context for the chapters comprising Section I. Section II analyses the context in more depth.

Chapter 1 concludes with a brief outline of the book.

What This Book Is About

This book explains the processes and best practices for designing integrated products, manufacturing systems and social systems (people)—4Ps and 1M—and harmonising the different design activities (Figure 1.1). Figure 1.1 is read by following the arrows; for example, *design (noun) is a model of a product*. **4PM** is a memory aid for what the book is about.

The word *design* has two meanings: design is used as a noun and as a verb. Design (noun) is a model of the structure of a product or system. Design (verb) is a process. Design (verb) creates a design (noun).

Design (verb) is the greatest ability humans possess. It is a process by which we imagine non-existent worlds and bring them into existence. It allows us to shape the world according to our thoughts. The range of things that can be designed covers all aspects of human life: clothes and accessories (fashion design), buildings (architectural design), building interiors (interior design, retail design and exhibition design), media (graphic design, multimedia design, branding and web design), jewellery (jewellery design), consumer goods (product design, packaging design, engineering design), industrial goods and complex systems (industrial design, engineering design), and social systems (service design, organisational design and socio-technical system design).

Our book is restricted to the design of consumer and industrial products, manufacturing systems for small to medium manufacturing companies (SMMCs) and the social systems to support design and manufacturing. We assume SMMCs design *and* manufacture their products.

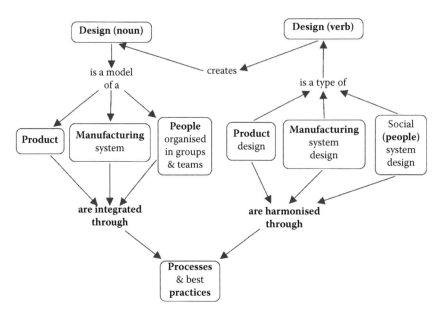

FIGURE 1.1
What this book is about.

Our book explains the two methods for achieving design synthesis: integration and harmonisation. Product, manufacturing system and social system architectures are *integrated* (united or combined to form a whole that is greater than the sum of the parts). The concurrent processes to design the architectures are *harmonised* (made compatible or coincident with one another).

Who This Book Is For

There are three main driving forces behind the book. The first is to present a text that covers both the technical and social aspects of design synthesis. The second is to present research and best practices that are no longer readily available, e.g., socio-technical systems approach. The third is to embed design synthesis within a framework of regenerative development. We strongly believe that designers have a social responsibility to ensure their products and systems are sustainable and will support the co-evolution of human societies and the environments within which they are embedded.

The key benefits of our book lie in its wide scope of coverage, depth of coverage, multi-disciplinary perspective and presentation of research and practices that are difficult or almost impossible to find. Our book is primarily intended for senior undergraduate or post-graduate engineering and technology students specialising in design and manufacturing, and designers, engineers and managers in SMMCs. We show how to:

- Design products to match manufacturing systems
- Design manufacturing systems to match products
- Harmonise concurrent product, manufacturing system and social system design
- Design, integrate and manage design teams
- Design and implement manufacturing systems that jointly optimise the technological and social systems

Readers can use our book to jointly optimise the design/re-design of new products and manufacturing systems to produce better quality products faster and cheaper.

An instructor's manual and PowerPoint slides are available for instructors. The manual provides teaching tips on how to use the book and present the PowerPoint presentations, and solutions to the exercises.

Improving Your Learning

Reflection is an important activity in designing and an aid for learning. We recommend you (the reader) start a reflective journal. You can record your reflections as you read the chapters, complete the exercises and apply the ideas in work and life.

What is a reflective journal? A reflective journal is a written record of your reflections on your activities as you learn. Each reflection is a purposeful, conscious effort to understand how your learning approach limits the effectiveness of your learning. The journal and the reflective process are intended to improve your learning effectiveness. Your journal is personal and unique. Format it to suit your personal style.

The steps to write a journal entry are:

1. Describe the significant event you are reflecting on. You may have difficulty understanding a concept. Or the difficulty may be in applying a concept. Or maybe there is no difficulty; you are simply looking at how to improve your learning. You should include events for which you have both negative and positive feelings, as you can learn from both successes and failures.

2. List the key assumptions relating to the significant event. The aim of this step is to get you to think about the assumptions you have made

and that you believe relate directly to the significant event. One effective way to do this is to create a concept map about the event.

3. Interpret the event in terms of whether it validates or invalidates your assumptions. Think about how you feel.

4. Evaluate your learning approach to determine how to improve its effectiveness by changing assumptions, modifying it, or by adopting a different approach.

5. Plan and execute the changes.

The reflective journal is one way of improving learning. Concept maps are another way. You have already encountered one concept map in this chapter (Figure 1.1). In the next section we explain how concept maps aid learning and describe how to create them.

About Concept Maps

According to Novak (2010, p. 23), "meaningful learning results when the learner chooses to relate new information to ideas the learner already knows". There are three requirements for meaningful learning:

1. Relevant prior knowledge: The learner has some knowledge that is related to the new ideas in some non-trivial way. We assume readers have background knowledge in science and engineering.

2. Meaningful material: The new material must present concepts that are related to each other.

3. The learner must choose to learn meaningfully: The learner must deliberately choose to assimilate the new knowledge with knowledge he or she already possesses and have the competence to do so. The learner comes to the learning situation with an organised body of knowledge. His or her task is to incorporate the new ideas into this body.

Novak argues that the major impediment to meaningful learning is the assimilation process (requirement 3). He has developed a tool to improve the effectiveness of assimilation: the concept map. A concept map is a diagram showing key concepts and their interrelationships. The map is usually ordered from top to bottom with more general, inclusive concepts at the top and more specific concepts at the bottom. The relationships between concepts are indicated as propositions. Each map has a focus, which can be indicated by a focus question. For example, consider the focus question: What is a concept map? Our map for this is shown in Figure 1.2.

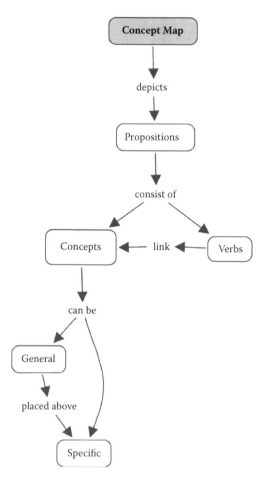

FIGURE 1.2
What is a concept map?

We use a focus concept in a grey box, instead of a focus question—this is our idea, not Novak's. The focus concept is usually shown at the top, but this is not essential, as you will see later in the book.

The first proposition shown is "concept map depicts propositions". It means what it says: concept maps depict (show or display) propositions. A simple proposition contains two concepts and a verb. The verb connotes the relationship between the concepts. The map shows simple propositions, which link two concepts only. Compound propositions can be constructed by linking several simple propositions.

Concept maps provide a general overview of the topic being discussed. Novak (2010) says learning is improved if general concepts are presented first, before specific concepts. For this reason, the chapters in the book begin with concept maps. Each map shows the key concepts and their relationships for that chapter and those of other chapters relevant to the particular chapter. The concepts from other chapters are highlighted by dashed boxes.

Our concept maps were produced using CmapTools version 5.04.02. CmapTools is free software from the Institute of Human and Machine Cognition (http://cmap.ihmc.us/). We recommend you download the free software and create your own maps as you work through the book.

The exercises at the end of this chapter are intended to help you reflect, develop skills in creating concept maps, and start your journal.

Introduction to Global Manufacturing

It is common knowledge that we live in a global world. But is it? Ghemawat (2011) is one of the few people who have carried out research to estimate the degree of globalisation. His results are astonishing:

- Only 17–18% of Internet traffic crossed national borders in 2006–2008 (p. 26).
- First-generation immigrants account for only 3% of the world's population (p. 27).
- Only 2% of students study overseas (p. 27).
- 90% of the world's population never leave the country in which they were born, not even for a holiday (p. 27).
- Trade intensity (a measure of exports and imports) peaked at 29% in 2008, then dropped to 23% in 2009 (p. 28).
- Global exports as percent of GDP is estimated at 20% (p. 28).
- Foreign direct investment (FDI) averages around 10% of world total investment (p. 29).

These are not the numbers you would expect to see if the world was truly global. The world is partially global and that is all. There are many impediments to the movement of products, people, information and capital across national borders. Governments restrict immigration, and impose trade and capital barriers. The recent global financial crisis has exacerbated the situation: governments have increased the barriers as they seek to protect jobs for their people and their economies from foreign competition, and investors are fleeing from the government bond markets.

You may be wondering whether it is feasible for SMMCs to compete globally. Aren't they too small to go global? The answer is no: size is not a barrier. Germany and the Scandinavian countries provide the proof. Germany is the world's second largest exporting nation and was the world's leading exporting nation until 2009, when it was overtaken by China. Germany exports more than the United States. *The majority of Germany's exports are from small and medium companies who dominate the international markets they are in.*

There are many challenges for global SMMCs. The biggest challenge is uncertainty. There are major changes taking place in world economies, politics and demographics. Economic power is shifting rapidly from the developed economies to the emerging and developing economies, especially the BRICK nations: Brazil, Russia, India, China and South Korea. Western economies are heavily in debt, whereas many emerging economies are not. Governments of developed economies have implemented austerity measures to reduce their debt burden. Jobless rates are high, especially amongst the young. SMMCs in these countries will struggle if they only serve local markets, and many will not survive.

Politically, there is a shift from a unipolar world dominated by the United States to a multi-polar world dominated by no nation. As America's economy declines, so will its global reach and power. Furthermore, all sovereign states are demanding an increased say in world affairs. They are not prepared to accept American hegemony. We are entering a new era where governments have no experience. Mistakes and miscalculations can be expected as America's political and military alliances unravel and emerging and developing nations challenge America's right to rule (Moran, 2012). There is a very high risk that regional conflicts will escalate into local or civil wars.

The global banking crisis hit many SMMCs very hard. SMMCs do not have a lot of liquidity. When the crisis hit, those that had borrowed money from banks found themselves in trouble. The money is commonly loaned against the stock market value of the company. The crisis led to a drop in the market value of all companies. Unfortunately, the banks have a rule that states the amount of the loan cannot exceed a certain percentage of the market value of the company. When the market dropped many companies exceeded the percentage cutoff. They either had to re-negotiate with the banks or raise money on the stock market at a time when investors were pulling out. Some were successful, some were not. Stock market uncertainty means that SMMCs are still vulnerable to this threat.

Investors are sitting on the sidelines at the time of writing (2012) not knowing where or when to invest. Bond markets have taken a beating. It is not all doom and gloom, though. Some investors are moving back into the corporate bond and equity markets. Raising capital for overseas ventures is a little easier now than it has been since 2008.

Social unrest is very high in many countries, developed and developing. This is a result of high and widespread unemployment and ruthless autocratic rulers. Riots and organised protests are common. Factories can be closed very quickly and supply chains cut. Riots and protests are not predictable but should be expected.

There is a major change in demographics in all societies. Populations are aging in the developed countries, and at the same time the number of working adults is shrinking. Countries that pay pensions through current taxes face two difficulties. The first is obviously financial. The tax base is shrinking whilst the pension base is growing. Governments cannot keep raising taxes. The second is: Who is going to look after all the old people? The current solution is to employ foreign workers from developing countries. However, as these economies grow in wealth, the workers will find they can earn more money staying at home. The current solution is not a viable option in the long run.

The demographic situation is the opposite in developing countries. They have a rapidly growing young population and an expanding base of working adults. The danger here is ensuring there are sufficient jobs for everyone. Therefore developing countries often demand that foreign companies employ local people. Typically locals do not have the technological education and skills SMMCs require.

There are further risks arising from management of global supply chains, scarcity of resources and environmental degradation. The risks for globalisation have never been higher at any time since the Second World War.

On the other hand, the opportunities have never been better either. New markets are opening up in emerging and developing countries as people gain wealth. There are also new market opportunities arising from the changing demographics, e.g., aged health care and retirement products and services. SMMCs have an advantage over multinational corporations (MNCs) as they are more nimble and can adapt more rapidly. Also being small, they do not attract attention in the same way MNCs do.

Book Outline

The concepts and processes presented in our book are complexly intertwined. The concept map in Figure 1.3 will help you grasp the key interrelationships. As you can see, the book is divided into four main parts:

- Section I: Foundation
- Section II: Context
- Section III: Synthesis—Technologies
- Section IV: Synthesis—People

Section I is the foundation for Sections III and IV. We recommend you read the four chapters comprising Section I first, although some concepts may be familiar to you. These chapters present old, familiar concepts in new ways and new ideas that are not well known.

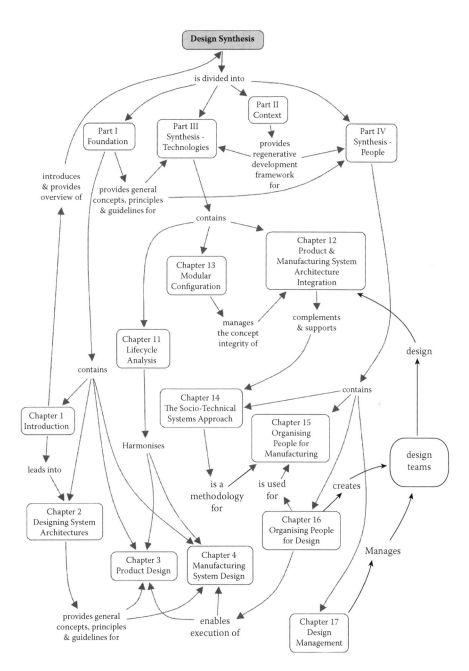

FIGURE 1.3
Concept map of the book.

This introductory chapter leads into Chapter 2, which discusses generic system concepts, practices and guidelines for creating and implementing design processes and designing system architectures.

The key topic of Chapter 3 is design (verb). We are the first to make a distinction between two different design processes: design as a personal activity and design as a phase. The distinction is critical for understanding the literature on design and executing design effectively in industry. Next we introduce you to batch design, a process for designing products so they can be made efficiently using batch manufacturing (Chapter 4). We conclude Chapter 3 by describing five strategies to develop products. They have a major impact on how product and manufacturing system architectures are integrated and how their design processes are harmonised.

Chapter 4 focuses on the most suitable type of manufacturing for SMCCs: batch manufacturing. We describe group technology, a methodology for designing part manufacturing cells. Design guidelines for assembly and test cells are also discussed.

Section II provides a context for designing products and systems (Sections III and IV) to support regenerative development. We propose a new model for evaluating the complex milieu within which global SMMCs operate. The model is called EASTEEP, an acronym for **e**thics, and **a**esthetic, **s**ocial, **t**echnological, **e**nvironmental, **e**conomic and **p**olitical contexts.

Section III shows how to match product family architectures with the technological system architecture that makes the products. The design processes for products and manufacturing systems are harmonised through intersecting development life cycles, discussed in Chapter 11.

Chapters 12 and 13 go hand in hand. Chapter 12 talks about modular product architectures, how to design them, and how technological system architectures can be designed to match modular product architectures. Chapter 13 describes configuration matrices, how to create them and their relevance for managing product configurations. The matrices can and should be developed during the design of modular product architectures because they are a powerful means for verifying the architecture and maintaining concept integrity.

Section IV shows how to:

- Integrate people with the technological system of a manufacturing system
- Design an effective organisational structure for a SMMC
- Organise people into design teams
- Integrate the design teams with the organisational structure of the SMMC
- Manage design teams

Section IV starts with a discussion on the socio-technical systems approach (Chapter 14). This is an old approach that has been largely forgotten today. Yet it is important because it provides a scientifically proven basis for designing work groups. The ideas discussed in Chapter 14 provide the foundation for designing the organisational structure of SMMCs presented in Chapter 15.

Chapter 16 shows how to organise people into design teams and how the teams can be interlinked. The design teams execute product design (Chapter 3) and manufacturing system design (Chapters 4 and 15) to create product architectures (Chapter 12) and manufacturing systems (Chapters 4, 12 and 15), respectively.

Design teams must be properly managed to ensure they deliver high-quality designs on time and within budget. Chapter 16 talks about project management, which manages schedules and cost, and risk management for managing risks. Chapter 17 discusses design management, which manages the quality of a design and design-related experience.

Exercises and Problems

1. You have read the introduction. Are you excited about this book or not interested? Record your feelings and examine why you feel the way you do.

2. Create a concept map based on your current knowledge to answer the focus question: What is design?

3. Create a concept map based on your current knowledge to answer the focus question: What is batch manufacturing?

References

Ghemawat, P. 2011. *World 3.0.* Boston: Harvard Business Review Press.

Moran, M. 2012. *The reckoning: Debt, democracy, and the future of American power.* New York: Palmgrave Macmillan.

Novak, J.D. 2010. *Learning, creating, and using knowledge.* 2nd ed. New York: Routledge.

2

Designing System Architectures

Introduction

A thorough understanding of systems theory, principles and practices is needed to design system architectures. Unfortunately, different disciplines have developed their own theories and practices independently. There is no universal system theory and practice accepted by all scientists. This would not be a problem if engineers could ignore political, social and environmental issues in the performance of their work. But in today's world they cannot. We face the same issues in writing this chapter. We have the difficult task of trying to be as inclusive as possible, yet at the same time reconcile the conflicting definitions and practices from different disciplines.

Complex engineering systems are developed using a methodology known as systems engineering (Blanchard and Fabrycky 2010, Buede 2009, Sage and Armstrong 2000, Shisko 1995). Systems engineering proved highly successful in the 1950s and 1960s when it was first applied to military and space projects. It failed miserably when applied to public systems like health, transportation and education. System researchers studied the reasons for these failures and developed new practices. Peter Checkland (1993, 1999) and his colleagues in the UK developed soft systems methodology. Russell Ackoff and his colleagues (1974, 1981, 1999a, 1999b) in the United States developed an approach called interactive planning. The new approaches sweep in social, political, moral and aesthetic issues that had been ignored by systems engineering. They embed systems engineering within a more inclusive framework. Systems engineering is discussed in this chapter. Interactive planning is described in Section IV.

Figure 2.1 shows the concept map for Chapter 2. The focal concept is *system*, whose meaning is explored in depth. We talk about a system philosophy based on the pursuit of ideals and show how systems can be designed to further progress toward ideals. We also discuss the design, validation and verification of system architectures.

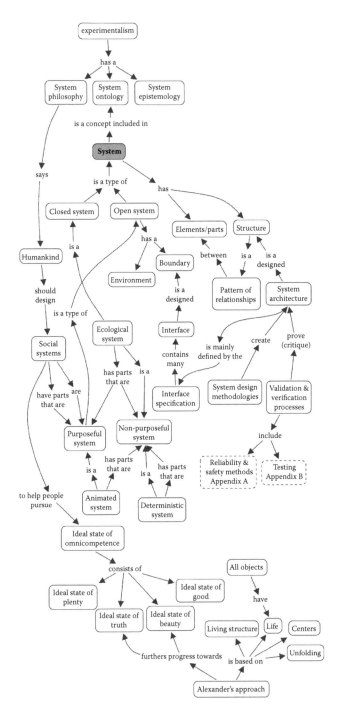

FIGURE 2.1
Concept map of Chapter 2.

Basic Concepts

System: A set of elements directly or indirectly related to each other.

The most important features of a system are (Ackoff 1999a, Beer 1981, Alexander 2002a):

1. Each element of the system is affected by or affects at least one other element in the set. The elements of a designed system usually exist before they are combined to form a system. On the other hand, in natural systems, the elements and the system are created at the same time, with the elements providing the conditions for the existence of each other. Therefore if one element dies, the other elements may also die and the system ceases to exist.

   The elements are commonly referred to as parts and sometimes as centres.

2. The whole set (the system) cannot be divided into independent sub-sets. When a system is taken apart, it loses the properties that identify it (these are called essential properties).

3. Each element has an effect on the properties of the whole set. When an element changes, the system's essential properties will also change.

4. The properties of the whole set are different from the properties of the elements. Systems often surprise us because they behave in ways we do not expect based on what we know about the elements. The reason we are surprised is because system behaviour is highly non-linear and we are not used to dealing with non-linearity. For example, hydrogen and nitrogen combine to form ammonia. Ammonia has a pungent smell, but hydrogen and nitrogen are odourless. "Pungent smell" is a system property that does not exist in the elements. The system properties are sometimes referred to as emergent properties to indicate that they are different from the properties of the elements, but also derive from the interactions between them.

5. The properties of the whole set affect the properties of each element.

6. The elements of the system are systems. Systems contain systems and are themselves contained within systems. The definition is recursive. The recursive definition and the complex nature of physical systems mean that complexity does not disappear when a system is divided into smaller and smaller sub-systems.

7. Natural systems are self-organising and evolve over time, so do many social systems. Man-made products and technical systems are not designed to evolve. They degrade over time unless they are maintained.

System structure: The pattern of relationships between the elements. A system structure is holistic because all elements of the system are involved in producing and maintaining it. Structures can change slowly over time, e.g., a bearing that wears out, or suddenly, e.g., when a shaft breaks. The characteristic system behaviour changes when the structure changes. In the short term, a structure remains the same. If it didn't we would not be able to identify the system at all. Natural systems can only achieve structural stability if they have dynamic processes actively working to prevent their structures from changing. This is called self-regulation. Self-regulation preserves structure. Self-organisation destroys structure because re-organisation is a change in structure.

System architecture: A system structure designed by people.

Environment of a system: Elements and their relationships that are not part of the system or its boundary, but which affect or are affected by the system.

Closed system: A system that has no environment.

Open system: A system that has an environment. Open systems are extremely complex. The complexity arises from the large number of elements, the large number of relations between the elements, the large number of relations between the system and its environment, and the inherent complexity of the system boundary, the elements and the relationships.

System boundary: An arbitrary line, area or volume that demarcates a system from its environment.

Purposeful system: An open system that can (1) produce the same function in the same environment using different courses of action and (2) produce different functions in the same and different environments. Typical examples are animals, people and social systems.

Ecological system: A closed system that has no purpose, but contains some elements that are purposeful systems. An ecology may contain a mixture of deterministic systems (e.g., river systems and forests), animals, people and communities. It is possible to model an ecology with one type of purposeful system; e.g., a social ecology consists solely of social systems.

System life cycle: A model of a repeatable process that is the complete life cycle of a system. The main phases of a life cycle for designed systems are specify, design, manufacture/construct, deploy/distribute/commission, operate, maintain and support, and retire (Chapter 11).

System Philosophy

Throughout history philosophers have discussed how people ought to live their lives. The "oughts" are expressed as ideals: states or outcomes that cannot be reached, but which can be approached indefinitely. Experimentalism (Britton and McCallion 1994) is a philosophy of life based on ideals. It is a modern development of the American pragmatist approach founded by Charles Pierce (see Turrisi 1997 for details about Pierce's philosophy). It started as a philosophy of science, in search for the "truth". The experimentalists followed the pragmatist tradition by assuming truth to be an endpoint (ideal) of science, not the starting point. Scientific progress is viewed as an infinite sequence of interlinked sets of facts and laws, each set more closely approximating the truth (the ideal). Neither facts nor laws are considered inviolate; both will be revised during progress toward the ideal.

One key notion of this approach is the assumption that all problems in science are interrelated; therefore all sciences are relevant in any particular scientific enquiry. Therefore the experimentalists extended their philosophy to include all sciences. This led to the discovery that they could not separate pursuit of the ideal of truth from other ideals people seek. Consequently, they broadened their philosophy and approach to include these ideals, and their philosophy became a philosophy of life.

Why should people pursue ideals? An inherent characteristic of human nature is our ability to desire. Desire is meaningless unless it can be achieved. Hence to desire anything we must also desire the ability to achieve our desires. The ability to achieve any and all desires must be unattainable, because if it were attainable there would be nothing left for others to strive for. The experimentalists argue it is the one ideal sought by all people of all times. It is an inherent characteristic of all humans, past, present and future.

Historically, the unlimited ability to achieve desires has been considered the ideal state of omnipotence, but for experimentalists it is taken as the *ideal state of omnicompetence*. The difference in meaning between these two conceptions is considerable. Omnipotence is the ability of one person or a small group of people (the powerful) to achieve all their desires by controlling other people (the powerless). The powerless people do not achieve their desires. Their sole purpose is to serve the powerful. It is obvious that omnipotence is not suitable as an "ideal of unlimited ability" because it does not include everyone. Omnicompetence, on the other hand, is a utopia where everyone has the unlimited ability to achieve their desires.

Although the concept of ideals may seem strange to engineers, we would like to point out that ideals have a long association with engineering. One of the fundamental cycles in thermodynamics is the ideal Carnot cycle. It is a hypothetical cycle that is impossible to achieve in practice, Yet it has played a major role in the development of more efficient thermodynamic engines, because it serves as a benchmark against which the actual performance of engines can be compared. The comparison of an actual cycle against the ideal helps engineers determine where they should focus their efforts to re-design engines to improve performance.

Ancient philosophers identified three ideals: the good, the beautiful and the truth. Modern philosophers added a fourth ideal: plenty. According to the experimentalists these four ideals are necessary and sufficient to achieve the overall ideal state of omnicompetence. The ideal state of beauty relates to the aesthetic aspects of life. The ideal state of good relates to the moral aspects of life. The ideal state of truth relates to knowledge and understanding. The ideal state of plenty relates to having resources available when needed and the ability to cooperate in using them, i.e., the economic and political aspects of life. The key characteristics of ideal-seeking systems are (Emery 1974, 1981):

1. Only individual people can be ideal-seeking systems. "By implication, institutions or organizations that men have created can be, at best, purposeful systems, no matter how old and sacrosanct they may be. They can purposefully act to create conditions under which more of their members can, on more issues, be ideal seeking systems" (Emery 1974, p. 65).

2. The ideal-seeking state can be sustained only temporarily. "As we have defined it the pursuit of the ideal is a pursuit of the infinite, the unattainable. For the individual this could produce only informational overload and nervous breakdown if he remained focussed on ideals, the choice of purposes. What does seem empirically established is that men support each other to be ideal seeking" (Emery 1974, p. 65).

3. Ideals emerge within group life. A hermit cannot pursue ideals.

4. Other ideals are sacrificed if one ideal is pursued single-mindedly. Pursuit of the four ideals needs to be harmonised.

5. "Deciding on what sacrifices of other ideals should be made in any particular choice between purposes is the essence of wisdom" (Emery 1974, p. 66). This statement includes selecting purposes that facilitate pursuit toward two or more ideals simultaneously.

Social systems are not ideal-seeking systems but can and ought to be designed to help people pursue ideals.

The four ideals provide a general framework for evaluating human behaviour, social systems and systems designed by people (designed systems). Ideal-seeking social systems can be designed using a methodology known as interactive planning (Ackoff 1981, 1999a). Alexander (2002a, 2002b, 2002c, 2002d) provides a conceptual basis for designing physical systems that further progress toward the ideals of truth and beauty. First, consider the ideal state of truth. The purpose of pursuing the ideal state of truth is to increase knowledge and understanding for the benefit of people living today and in the future. In terms of benefits, Alexander argues that systems should be designed to enhance and improve the ecologies within which they co-exist. This is achieved by giving them more life.

Second, consider the ideal of beauty. Pursuit of this ideal measures the aesthetic progress of designed systems, whether they are becoming more beautiful or ugly. Alexander argues very strongly that 20th-century architectural practices produce environments that are ugly. Their ugliness derives from their lack of life. He proposes a new approach to architecture that will improve the life of designed systems and natural environments, making them more beautiful. Systems with more beauty also have more life.

Alexander's approach is based on the following concepts: life, living structure, centres, wholes and unfolding. He assumes that "human feeling is mostly the same, mostly the same from person to person, mostly the same in every person" (2002a, p. 3). He provides strong evidence that people evaluate art, artefacts, nature, and architecture in the same way. The evidence consists of the results of experiments in which paired images of art, artefacts, buildings, etc., are shown to people and they are asked to select the image with more life in each pair. With few exceptions people select the same image of each pair as having more life. These studies have been conducted over many years with large numbers of people, with the same results. Hence it is an inherent human ability.

Alexander infers from this that *all* physical objects have some degree of life, even objects considered to be inanimate, like a rock or metal window frame. In his own words:

> What we call "life" is a general condition which exists, to some degree or other, in every part of space: brick, stone, grass, river, painting, building, daffodil, human being, forest, city. And further: The key to this idea is that every part of space—every connected region of space, small or large—has some degree of life, and this degree of life is well defined, objectively existing, and measurable. (Alexander 2002a, p. 77)

His view has a sound psychological basis. Our feelings are our evaluations of the world around us. Objects with more life generate a greater feeling of well-being than those with less life. Another way of stating this is to say objects with more life improve our quality of life, whilst those with less life detract from it. Alexander (2002a) expresses this eloquently:

> A place which has the deepest life is one in which I reach a deeper level of life inside my self, and in my spirit. The quality of life I attain—its depth—in any given building is part of the way I experience that building. (p. 61)

For Alexander, physical space is a system, whose elements are centres. A centre is an ordered volume of space that focuses our attention.

The centredness of the centre is relative to other parts of space, which are also centres. Hence the centres affect the properties of each other and provide the conditions for each other's existence. Each centre also affects the wholeness of the set of centres, which is an emergent property of the system of centres. And the wholeness affects the properties of the centres. Furthermore, each centre is itself a system of centres; the definition is recursive.

The concept of centre is illustrated in Figure 2.2. We start with a blank piece of paper (Figure 2.2a). The whole piece is a centre and there are four corners that are also centres. Now a dot is placed in the middle of the paper (Figure 2.2b). The dot is a centre. As a focus of attention the dot has a halo around it (Figure 2.2c); this is another centre. The dot divides the paper into two vertical rectangles (Figure 2.2d), which are centres. Similarly it divides the paper into two horizontal rectangles (Figure 2.2e). There are four rays emanating from the dot vertically and horizontally to form four ray centres (Figure 2.2f). The four rays form a white cross, which is another centre (Figure 2.2g) and in so doing creates four smaller rectangles. Similarly, there are four rays from the dot to the four corners of the piece of paper, the diagonals (Figure 2.2h). Taken separately, each diagonal forms two triangles as shown in Figure 2.2i and j. Together they form a diagonal cross and four triangles (Figure 2.2k). In total, over 20 centres are formed by the placement of a single dot in the middle of the paper.

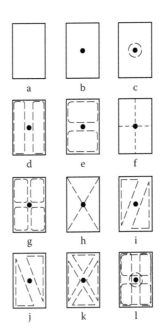

FIGURE 2.2
Illustration of centres and system of centres. (Adapted from Alexander, C., *The Nature of Order. Book 1: The Phenomenon of Life*, Berkeley, CA: Center for Environmental Structure, 2002a.)

One key point to note is that centres often overlap each other, as illustrated in Figure 2.2l for the four rectangles and the halo. This feature of centres is significantly different from the concept of parts, which are assumed to have distinct boundaries that do not overlap. For example, a piston pump and a valve in a hydraulic circuit are treated as separate and distinct entities whose boundaries do not overlap. Their boundaries stop at the points where they attach to the tube connecting them. However, the two parts can overlap through vibration. The pump sends a pulsating flow through the hydraulic tubes and the flow may cause the valve to vibrate, affecting its performance. In effect, the boundary of the pump is overlapping the boundary of the valve through the induced vibrations.

Electromagnetic fields created by electrical/electronic components also overlap. During the design of electrical/electronic systems this interference must be taken into account as it can affect their performance. Other forms of overlap include heat, noise and light. Designers should think of engineering components as centres rather than parts to ensure overlap is considered during component design and not left for a system engineer to resolve during integration. In this book, we will use the term *part* to refer to an element of a physical system, due to its common usage in engineering. However, you are reminded that when you read the term *part* we really mean *centre*.

Each centre has some degree of life that can be measured objectively. It follows therefore that a system of centres also has some degree of life. Alexander uses the term *living structure* to refer to the structure of a system of centres. The life of the living structure and the wholeness of space are produced by the centres and their configuration. The centres and the system of centres in natural systems are created together at the same time and evolve together. The architecture of buildings and communities can be created in the same way. During the evolutionary process, each new structure builds on the previous structure but is also different from it. New centres will be formed and previous (old) ones may be destroyed. Alexander calls this process *unfolding*.

A very simple illustration of an unfolding process is shown in Figure 2.3. We start with a blank piece of paper representing a room of a house (Figure 2.3a). A dot is placed in the middle of the paper (Figure 2.3b). The long side of the room opens to a garden (on the right), so the architect decides to have a lounge/dining room facing the garden, and selects the centres shown in Figure 2.3c. Two alternative solutions will now be briefly described. In the next step, shown by the solution on the left, the architect decides to focus on the large rectangle and removes the centre dot (Figure 2.3d$_1$). In the final stage for this development, the architect places a dot in the lower half of the large rectangle to designate a kitchen cum dining area (Figure 2.3e$_1$).

An alternative development is shown on the right. Here the architect decides to focus on the halo surrounding the dot in the centre, and adjusts the three rectangles accordingly (Figure 2.3d$_2$). Finally in Figure 2.3e$_2$, the

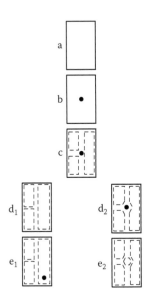

FIGURE 2.3
Illustration of unfolding.

dot is changed to a hexagonal shape, representing a fountain, to accentuate the centre of the room. Further progress of the two developments would lead to two totally different designs. However, in each stage the development proceeds by building on or removing centres from the previous stage.

This is an extremely brief description of a complex topic. Living structure cannot be understood and appreciated by reason alone (the knowledge domain). Other human faculties are required. One needs to experience and evaluate feelings generated by images of life or life itself; that is, it is necessary to develop skills in the affective domain. We recommend you read Alexander's book (2002a) *The Nature of Order. Book 1: The Phenomenon of Life* to gain and develop an affective appreciation of living systems.

A sustainable lifestyle is implied in the definitions of the four ideals: plenty, good, truth and beauty. Progress toward a sustainable lifestyle is measured as progress toward the four ideals. Ideals provide a means for assessing designs. Typical questions that can be asked of any design include the following. Does a design have more life or less life than other alternative designs? Is it more beautiful or more ugly? Does it improve the life of the surroundings within which it operates or detract from the life? Designs can be improved using Alexander's concepts of living structures, centres, wholeness and unfolding. His work is general in that it covers both designed and natural systems. Social systems can be designed to help people pursue ideals using interactive planning (Chapter 15).

Describing Systems

An open system is defined by four types of interdependencies (Emery 1981), as depicted in Figure 2.4 (subscript 1 refers to the system and subscript 2 refers to the environment):

- L_{11}: Interdependencies between elements contained wholly within the system. These define the system's characteristic way of behaving.
- L_{12}: Interdependency produced by a process that starts in the system but ends in the environment; i.e., it crosses the system boundary from the inside. It defines what the system can do to the environment. This kind of interdependency is referred to as an output in engineering.
- L_{21}: Interdependency produced by a process that starts in the environment but ends in the system; i.e., it crosses the system boundary from the outside. It defines what the environment can do to the system, i.e., the system's responsiveness to changes in the environment. In engineering, this kind of interdependency is referred to as an input.
- L_{22}: Elements or processes that are contained wholly within the environment. These affect the system indirectly through the L_{12} and L_{21} processes.

The definition is completely general, covering all types of open systems.

Figure 2.4 shows a system at one level of analysis. There is a limit to the amount of detail that can be shown at one level. Therefore large systems

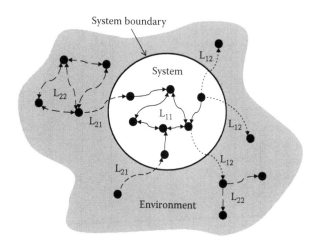

FIGURE 2.4
General model of an open system.

are decomposed into sub-systems, which in turn are decomposed into sub-sub-systems, and so on; resulting in a system hierarchy. The hierarchy is formed by the containment principle: each system level contains all the sub-systems below it in the hierarchy. The lower levels are more detailed views of the higher levels. The reverse is also true. Any system can be considered an element of a larger system, which is an element of an even larger system, etc.

One very efficient way to model a hierarchy is to use recursion (Ackoff 1981, Beer 1979). Recursion means invariant pattern. That is, the pattern between any two levels is the same for all level pairings, though the actual details will vary. The advantage of this kind of modelling is that if you know the pattern between any pair of levels, then you know the pattern for every other pair.

There are two kinds of architectural models of systems and system hierarchies: a *function* or *concept architecture* that shows the major functions or concepts and their interconnections or interrelationships, and an *embodiment (physical) architecture* that shows the major physical sub-systems and their interconnections. Both descriptions describe the structure of systems, but they do so in different ways. Functional flow diagrams, linked N^2 diagrams, layer diagrams, tree diagrams, interaction matrices and topological graphs are common representation techniques for function architectures. Flow diagrams, tree diagrams, network graphs, schematic diagrams, assembly diagrams, design matrices, cross-relationship charts and layout plans are representation techniques for embodiment architectures.

Systems can be classified into four different types (Ackoff 1999b), as illustrated in Figure 2.5.

Deterministic systems have no purpose and neither do their parts. "Their behaviour is completely determined by their structure, causal laws, and their environments if they are open systems" (Ackoff 1999b, p. 28). Typical examples include mechanisms, computers and plants.

Animated systems have purposes of their own, but their parts do not. The parts of animated systems are deterministic. Typical examples are animals and people. Social systems have purposes of their own and so do their parts (animated and social systems).

Ecological systems have no purpose, but contain parts that do (animated and social systems). They are closed systems that are composed of open systems and their environments; hence the general open system model can be used to model them. The open systems interact directly with each other or indirectly through their environments.

The four types of systems form a hierarchy as follows. Ecological systems contain animated or social systems, and may also contain deterministic systems. Social systems contain animated systems. Animated systems contain deterministic systems. There are two critical points to note. First, all designed systems are part of an ecological system whether this is recognised

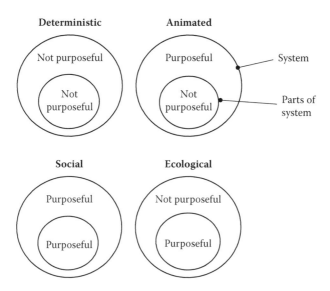

FIGURE 2.5
Types of systems.

or not. Their modelling and design should take into account their impact on the different types of systems that may be contained in the ecological system and on it as a whole.

Second, models of one type of system are often applied to systems of a different type, with serious adverse consequences (Ackoff 1999b). It is important to correctly identify the type of system being studied and describe it with a model appropriate to its type. For example, biological or organic analogies are not appropriate to social systems. Mechanistic homomorphisms and isomorphisms are only appropriate for deterministic systems. They do not apply to animated and social systems.

Describing Boundaries

A boundary (of a system) is an arbitrary line, area or volume that demarcates a system from its environment. In order to define a system it must be possible to determine whether an element is part of it or not. This is achieved by specifying an imaginary border around the system. Everything inside the border is part of the system. Many system scientists also include the border as part of the system, but this is not essential. It can be considered as a separate entity or as belonging to both the system and the environment. Everything outside the border is either not relevant (closed system viewpoint) or part of the environment (open system viewpoint). The same approach is used to determine the boundaries between the sub-systems of a system.

Why is the boundary arbitrary? The answer is that in reality boundaries are complex. They can be diffuse and deeply penetrate the system and its environment. Imagine you have been asked to define the boundary of an island. The government owning the island wants the boundary as large as possible as it affects the fishing rights for the surrounding sea. How would you do this? The simple solution is to say the boundary will be based on low tide. However, the tide changes throughout the year and over the years. Which low tide will you use? Will it be the lowest in the last year, last 10 years or last 100 years? Assume agreement has been reached on these points and that it has been decided the shoreline will be determined from an aerial photograph of the island. The next issue is the level of resolution to use. As the resolution increases the length of the shoreline will increase. In fact, a shoreline is a fractal boundary and its length is determined by its fractal dimension for a pre-specified level of resolution. So the "simple case" of determining the boundary of an island depends at least on arbitrary decisions about which low tide to use and the level of resolution.

Different disciplines have developed different techniques for defining and naming system boundaries. System engineers determine boundaries by cutting across the weakest links between the system and its environment and between its connected sub-systems. They call the boundaries interfaces (Grady 2006) and define them in an interface specification. An overview of the connected systems and their relationships can be shown using an N^2 diagram, cross-relationship chart or schematic diagram.

The boundaries for psychological and social systems are determined by a wide range of criteria, such as legal/regulatory, racial, ethnic, religious, cultural and political. We use a 50% rule for determining boundaries in supply chains. The rule is: if a company supplies 50% or more of its products or services to one customer, then that company is part of the customer's system. The reason for this is the customer controls the company, even though it may be commercially registered as a separate company. The customer has the market power to set price, timing of orders and quantity to produce. It may even specify how the products are to be manufactured. The supplier cannot refuse any demands from this customer because it cannot afford to lose his or her business.

A system interacts with the environment through its boundary. Sub-systems interact with each other through their boundaries. The interactions are achieved through physical connections, such as wires, tubes, air, etc. The interactions themselves may be physical, psychological or social. A connection has three components: two endpoints and a link between them, as illustrated in Figure 2.6.

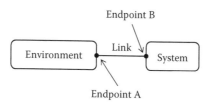

FIGURE 2.6
Components of a connection.

There are five commonly used ways to allocate the components to the environment, boundary and system:

1. Endpoint B is allocated to the system, and the link and endpoint A are allocated to the environment.
2. Endpoint B and the link are allocated to the system, and endpoint A is allocated to the environment.
3. The whole connection is allocated to the system.
4. Endpoints A and B are allocated to the environment and system, respectively, and the link is allocated to the boundary.
5. The whole connection is allocated to the boundary.

Describing Environments

There are five types of environmental influences (Grady 1998): natural environment, hostile environment, non-cooperative environment, cooperative environment and induced environment. The natural environment is described in terms of physical properties such as temperature, pressure, wind, rain, etc. There are many standards available that define the properties of the earth, its atmosphere and the space environment. Normally the environmental properties are specified independently of each other. However, some important aspects of the natural environment cannot be treated this way. They have to be considered as systems, e.g., the carbon cycle, water cycle and ecological systems.

The induced environment consists of those influences that exist only because the system exists, e.g., pollution. The hostile environment consists of systems whose intention is to prevent the system being designed from achieving its goals and objectives. It includes competitors, criminals and activists.

The non-cooperative environment consists of systems that are not intended to prevent the system from achieving its goals and objectives, but do so unintentionally, for example, a customs system that takes a long time to clear a cargo of goods, resulting in longer lead times and higher costs.

The cooperative environment consists of systems whose intention is to help the system achieve its goals and objectives, e.g., the customers of a system and companies in the supply chain. Customers support a supplier by giving money, advertising (telling others about their experiences) and providing feedback. Unfortunately customers can suddenly switch from cooperative to hostile if they are not happy with a supplier's espoused goals and objectives, or for reasons that have nothing to do with the supplier. Chinese netizens called for a boycott of Carrefour in 2008. They were protesting against an incident where pro-Tibetan protestors tried to prevent the Olympic torch, for the Beijing Olympics, being paraded through Paris. They targeted Carrefour because of its French origins.

The environmental aspects affecting the system are defined in an environmental specification (Grady 2006). It lists the environmental requirements the system is expected to withstand, for example, maximum and minimum temperatures. Considerable savings can be made by using existing standards for the environmental specification and existing test procedures and tailoring them to correspond to the actual conditions to which the system is expected to be exposed. The tailoring process determines the design requirements and the test procedures for environmental testing of the actual system.

A system may operate in more than one environment. The sequence in which it encounters the different environments can be described using an operational scenario. Each environment in the scenario will be specified by a set of requirements.

A system architecture may consist of a hierarchy of systems. There will be a hierarchy of environments corresponding to the different system levels. One of the major challenges in developing an environmental specification is determining how to combine the environmental requirements from the different levels into concise specifications for the system and its sub-systems in order to minimise design verification costs.

The parts of the environment affected by the system are defined by an environmental impact assessment or life cycle analysis. An environmental impact assessment is a report describing the impacts, positive and negative, that a system will have on the environment. The assessment is usually applied to manufacturing systems and other types of facilities. Studies of this kind are normally limited to the environment immediately surrounding the facility. The impacts of industrial products are defined by life cycle analysis. A full life cycle analysis is called a cradle-to-grave approach. It starts with the extraction of raw materials and ends when a system is retired. If the materials from a system are recycled or re-manufactured, then a new cycle begins, starting with the retirement phase of the original system. Life cycle analysis can also be applied to manufacturing systems.

Designing System Architectures

There are five main methodologies for developing architectures (Maier and Rechtin 2002): normative, rational, participative, heuristics and patterns.

Normative (solution based): A normative methodology codifies successful past experience for future use, for example, engineering codes of practice and engineering standards. Codified experience is specific to an industry and type of system. The key advantage of this approach is that past experience is recorded and available for future use, improving the success of new architectures and reducing the risk of failures. One major disadvantage of normative methods is they always lag behind current best practice. Furthermore, they are often produced by negotiation between the main parties affected by them, resulting in a compromise.

Rational (method based): This method uses science and technology to develop rules and methods for designing architectures. The rules and methods are validated by experiments and testing.

Participative: A participative methodology involves some or all of the stakeholders in the architecting process. The aim is for the stakeholders to input their viewpoints and values throughout the design process. The stakeholders become co-designers. Participative methodologies include interactive planning, soft systems methodology (Checkland 1993, 1999), socio-technical systems approach and agile software development (Shore and Warden 2007). Interactive planning and the socio-technical systems approach are discussed in detail in Section IV.

Heuristics: A heuristic is an experience-based guideline, for example, KISS—"keep it simple stupid". Maier and Rechtin (2002) provide a long list of heuristics for designing architectures. The major problem with heuristics is selecting an appropriate one to apply in a given situation. This in itself requires experience.

Patterns: A pattern is a partial architectural design that can be used over and over again in different situations. The concepts of patterns and pattern language were first expounded by Christopher Alexander and colleagues (1977, 1979). Patterns were quickly adopted by software developers and are now widely used to design software (Gamma et al. 1995).

Verifying System Architectures

System engineers use the terms *validation* and *verification* to refer to processes that prove a system will satisfy customer needs. Unfortunately, there is no general agreement on the meaning of these two terms. The following definitions are based on Grady (1998) and Shisko (1995). Validation is an assessment process to prove the right system is being built. It demonstrates that requirements are correct, understood and can be achieved with current technological capability. Validation is usually carried out before design.

Verification is an assessment process to prove that the system as built satisfies (complies with) the design requirements. Verification is carried out after design. In situations where the customer's needs are ill defined or uncertain, it is often better to validate by designing and building a prototype and asking the customer to directly evaluate this. Feedback from the customer is used to re-design the prototype or design a new one. In these situations design is carried out before validation, and validation and verification are performed at the same time.

Some commonly used validation techniques are:

1. *Market feasibility* (proving requirements are necessary): Quality function deployment (Cohen 1995), focus groups (Kreuger and Casey 2000) and idealised designing (Ackoff and Vergara 1981, Ackoff et al. 2006).
2. *Technical feasibility* (proving a system can be built to requirements): Design rules and guidelines, prototype development and demonstration, and design reviews.
3. *Environmental feasibility* (proving a system has minimal impact on or will not harm the environment): Environmental impact assessment and life cycle analysis.

There are seven main types of verification methods (Shisko 1995):

1. *Similarity*: Show that the current design is similar to a past design that has already been qualified to equivalent or more stringent requirements.
2. *Analysis*: Use of theoretical or empirical evaluation based on scientific or engineering principles, using accepted techniques. Design analysis methods can be divided into two groups: functional performance methods and reliability and safety methods. Functional performance methods include mathematical techniques and computer-based numerical techniques, e.g., finite element analysis (FEA), computational fluid dynamics (CPD) and electronic data automation (EDA). Reliability and safety methods include, but are not limited to, failure modes, effects and causality analysis (FMECA), worst-case analysis (WCA), fault tree analysis (FTA), safety analysis and sneak analysis. For a brief description of these techniques refer to Appendix A.
3. *Demonstration*: Demonstrate operational performance and compliance to requirements using the final system.
4. *Simulation*: Demonstrate performance and compliance to requirements using a prototype.

5. *Examination* (inspection or measurement): Visually inspect or measure physical characteristics of the system, with or without the use of measuring equipment.

6. *Test*: Measure performance and functionality under simulated or stimulated environments. There are two main test approaches. The *test-to-pass* approach is used to show compliance with the requirements. These tests simulate the normal environmental and operational conditions a product will be exposed to during its lifetime. The product is expected to perform its functions to the required standard under these conditions. It either passes or fails the test. The *test-to-fail* approach is used to provide assurance that the system is reliable (system can keep operating without failing) and durable (system can withstand wear-out mechanisms). These tests are carried out under simulated conditions (environmental testing) or stimulated conditions (accelerated stress testing). Testing is discussed in more detail in Appendix B.

7. *Review of records*: Assess manufacturing records to prove the system as built complies with requirements.

Verification is performed at different system levels to confirm that integration has been successful. It is normally performed in a bottom-up approach.

Verification starts in the design phase and continues to the retirement phase. *Design verification* starts during the design phase and continues into manufacturing. Design verification testing is referred to as qualification testing. *Manufacturing verification* starts in manufacturing and continues into deployment/distribution. Manufacturing verification testing is referred to as acceptance testing. *Operational verification* activities may be needed to confirm the performance of the system in actual use. *Retirement verification* proves compliance to disposal requirements.

Exercises and Problems

1. Imagine a household with two parents and two children living in an apartment in a city. Assume the household is a system.

   a. Draw a simple block diagram (one block only) of the system to show the major physical inputs and outputs.

   b. Which input is the most important? Explain your answer.

   c. List the main systems in the environment producing the inputs and those that are affected by the outputs. Classify them according to Grady's classification.

2. Walk around the campus, town or city you live in. Take a photo of a building that you think is particularly ugly. Take a photo of a

TABLE 2.1

Incidence Matrix for a Forest

	Sun	Hunters	Timber	Plants	Litter	Herbivores	Primary Predators	Decomposers
Sun	0	0	0	0	0	0	0	0
Hunters	0	0	0	0	0	1	1	0
Timber	0	0	0	1	0	0	0	0
Plants	1	0	0	0	0	0	0	1
Litter	0	0	0	1	0	0	0	0
Herbivores	0	0	0	1	0	0	0	0
Primary predators	0	0	0	0	0	1	0	0
Decomposers	0	0	0	0	1	1	1	0

building that you think is beautiful or pleases you a lot. Place the two photos side by side. Which photo has more life?

3. Table 2.1 shows the incidence matrix for a predator–prey system called forest.

   The matrix shows the relationships between the elements. A 0 means there is no relationship. A 1 means there is a relationship. The direction of the relationship is from column to row. For example, sun is related to plants, and the direction is from sun to plants.

   a. Draw a system diagram for the forest.

   b. Draw the system boundary.

   c. Is the system open or closed?

   d. Using Ackoff's classification, what type of system is it? What types are the elements?

4. Why is systems thinking important for professional engineers?

5. You are a member of a student team building a racing car to enter either Formula Student or Formula SAE (select the competition most appropriate to your location). Your task is to prepare a verification plan. List the validation and verification activities that should be carried out to prove the car. Explain the rationale for your answer.

References

Ackoff, R.L. 1974. *Redesigning the future: A systems approach to societal problems.* New York: John Wiley.

Ackoff, R.L. 1981. *Creating the corporate future.* New York: John Wiley.

Ackoff, R.L. 1999a. *Re-creating the corporation.* New York: Oxford University Press.

Ackoff, R.L. 1999b. *Ackoff's best*. New York: John Wiley.

Ackoff, R.L., and Vergara, E. 1981. Creativity in problem solving and planning: A review. *European Journal of Operational Research*, 7, 1–13.

Ackoff, R.L., Magidson, J., and Addison, H.J. 2006. *Idealized design*. New Jersey: Prentice Hall.

Alexander, C. 1979. *The timeless way of building*. New York: Oxford University Press.

Alexander, C. 2002a. *The nature of order. Book 1: The phenomenon of life*. Berkeley, CA: Center for Environmental Structure.

Alexander, C. 2002b. *The nature of order. Book 2: The process of creating life*. Berkeley, CA: Center for Environmental Structure.

Alexander, C. 2002c. *The nature of order. Book 3: A vision of a living world*. Berkeley, CA: Center for Environmental Structure.

Alexander, C. 2002d. *The nature of order. Book 4: The luminous ground*. Berkeley, CA: Center for Environmental Structure.

Alexander, C., Ishikawa S., and Silverstein, M., with Jacobsen, M., Fiksdahl-King, I., and Angel, S. 1977. *A pattern language*. New York: Oxford University Press.

Beer, S. 1979. *The heart of enterprise*. Chichester: John Wiley.

Beer, S. 1981. *Brain of the firm*. 2nd ed. Chichester: John Wiley.

Blanchard, B.S., and Fabrycky, W.J. 2010. *Systems engineering and analysis*. 5th ed. Englewood Cliffs, NJ: Prentice-Hall.

Britton, G.A., and McCallion, H. 1994. An overview of the Singer/Churchman/ Ackoff school of thought. *Systems Practice*, 7(5), 487–521.

Buede, D.M. 2009. *The engineering design of systems: Models and methods*. 2nd ed. New York: Wiley.

Checkland, P. 1993. *Systems thinking. Systems practice*. Chichester: John Wiley.

Checkland, P. 1999. *Soft systems methodology: A 30-year retrospective*. Chichester: John Wiley.

Cohen, L. 1995. *Quality function deployment*. Reading, MA: Addison-Wesley.

Emery, F.E. 1974. *Futures we're in*. Canberra: Centre for Continuing Education, Australian National University.

Emery, F.E. 1981. The emergence of ideal-seeking systems. In F.E. Emery (ed.), *Systems thinking*, 431–458. Vol. 2. Harmondsworth, UK: Penguin.

Gamma, E., Helm, R., Johnson, R., and Vlissides, J. 1995. *Design patterns*. Reading, MA: Addison-Wesley.

Grady, J.O. 1998. *System validation and verification*. Boca Raton, FL: CRC Press.

Grady, J.O. 2006. *System requirements analysis*. Burlington, VT: Academic Press.

Kreuger, R.A., and Casey, M.A. 2000. *Focus groups*. 3rd ed. Thousand Oaks, CA: Sage Publications.

Maier, M.W., and Rechtin, E. 2002. *The art of systems architecting*. 2nd ed. New York: CRC Press.

Sage, A.P., and Armstrong, J.E. 2000. *Introduction to systems engineering*. New York: John Wiley.

Shisko, R. 1995. *NASA systems engineering handbook*. Washington, DC: NASA.

Shore, J., and Warden, S. 2007. *The art of agile development*. Sebastopol, CA: O'Reilly.

Turrisi, P.A. (ed.). 1997. *Pragmatism as a principle and method of right thinking by Charles Sanders Peirce. The 1903 Harvard Lectures on Pragmatism*. New York: State University of New York Press.

3

Product Design

Introduction

Design is used as a noun and a verb. Design (noun) refers to the product or system architecture and the detailed form of the parts. Design (verb) refers to design as a process. Where there might be confusion between the two terms, we will qualify the term, as in the next sentence.

The concept map for this chapter is shown in Figure 3.1. As you can see, this chapter is the key linking chapter for the book. Not surprising for a book on design! The focus question is: What is design? The focal concept for the chapter is design (verb): a generic process applicable to product design and manufacturing system design. Product architectures—design (noun)—are discussed in Chapter 12.

Design (verb) is used to refer to three different processes: the product development process, a phase of the product development process and a personal activity. Pugh (1991) uses the term *total design* as a name for a product development process that includes the following phases: user need, design specification, conceptual design, detail design, manufacture and sales. Pugh says products are designed by teams of people from different disciplines. The disciplines produce partial designs that are combined to create the final design. The purpose of his book is to provide a framework for integrating the partial designs. It is useful from that viewpoint, but we see no need to rename the product development process. We do not recommend this usage of design.

Design as a personal activity produces a model of an object or system that does not exist and communicates this model to at least one other person. *Design as a phase* is an essential part of the product development process. Design as a phase includes two major activities: design as a personal activity and critique. A design is not complete until a critic has critiqued it. In art, the critique is normally performed once the design is completed. In engineering, design and critique may iterate many times before a design is finalised. Engineering critique is referred to as validation and verification (Chapter 2).

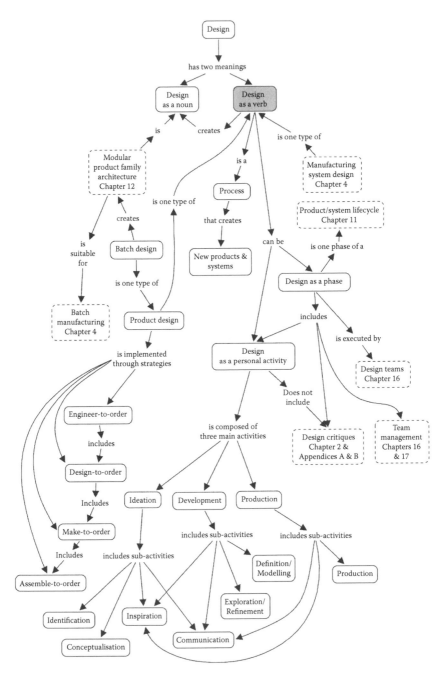

FIGURE 3.1
Concept map of Chapter 3.

Design as a Personal Activity

A person designs when he or she produces a model of an object or system that does not exist, but is capable of being brought into existence, and communicates the model to at least one other person. The model represents the essential properties of the object or system (the product). The set of properties is the design (noun). Many different types of models are used to communicate designs. Some of the more common ones are physical models, symbolic models, concept maps, concept boards, mood boards, textile and fabric boards, sketches, illustrations, drawings, plans, diagrams, charts, matrices and graphs.

The discussion in this section is oriented toward product design; however, design as a personal activity also applies to the design of manufacturing systems.

A designer may generate three different descriptions of products: conceptual or functional design, embodiment or preliminary design, and detailed design. A conceptual design describes the functional architecture, an embodiment or preliminary design describes the physical architecture, and a detailed design describes the shape and form of individual parts.

Design synthesis is the most important feature of design as a personal activity. It is a process of bringing together or connecting. Synthesis can be applied in different ways during the design process. It can be used to generate a design problem, search for and explore the design solution space, and generate design solutions (Chakrabarti 2002).

Synthesis is most often associated with creativity. Creativity is a mental process of divergent thinking. Divergent thinking occurs when two completely different things are identified as the same based on one property or because their names have similar sounds (Arieti 1976). An example of the former would be identifying a red car with a red rose because they are both red. An example of the latter is identifying a cog with a frog because they sound similar. Divergent thinking is based on simile, metaphor and analogy. A number of techniques have been developed to stimulate divergent thinking: brainstorming (Osborn 1963), synectics (Gordon 1961, Prince 1970), lateral thinking (De Bono 1983) and idealised designing (Ackoff and Vergara 1981, Ackoff et al. 2006).

Personal creativity can be cultivated through:

- Aloneness: Creativity is stimulated when a person is alone without being disturbed (Arieti 1976, Csikszentmihalyi 1997). Designers should develop habits of being alone and allow time in their day for quiet reflection on the design problem.
- Inactivity (inspiration): Designers should find and appreciate different sources of inspiration in everyday life and reflect on these quietly (Arieti 1976, Aspelund 2010).

- Observation and curiosity (Arieti 1976, Csikszentmihalyi 1997): Ideas can come from anywhere. Designers need to train themselves to be observant and curious. Csikszentmihalyi (1997) provides some good tips on how to do this.
- Creativity-stimulating techniques: Designers should learn the techniques for stimulating divergent thinking and apply them regularly in everyday life.

Design synthesis can also be achieved using rational thought processes and systematic methods. Systematic techniques for generating concepts include mind maps (Buzan and Buzan 1996), concept maps (Novak 2010), gap and attribute analysis (Cather et al. 2001), morphological analysis (Ullman 2009, Zwicky 1969), TRIZ (Rantanen and Domb 2007), extremes and inverses (Ullman 2009) and SCAMPER (Eberle 2008). Pahl et al. (2007), Hubka and Eder (1988), and Suh (1990) have developed systematic methods for designing engineering systems.

Design as a personal activity is a journey from the world of imagination to the world of objects (Aspelund 2010). Three purposeful activities composed of seven sub-activities must be carried out to successfully complete the journey (Figure 3.2). The purpose of ideation is to create the design brief and design concept. The purpose of development is to explore, refine and define the design concept. The purpose of production is to produce a physical prototype of the design for critique. There is no set sequence for conducting design as a personal journey. In some situations the activities may be carried out simultaneously; in others they might be performed in a fixed sequence. In all cases, there will be iteration.

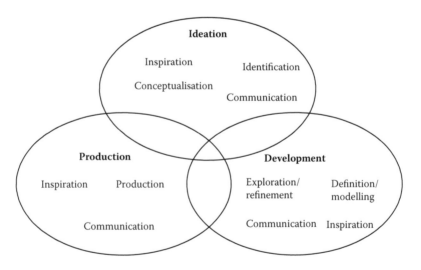

FIGURE 3.2
Design as a personal activity.

The seven sub-activities are inspiration, identification, conceptualisation, exploration/refinement, definition/modelling, communication and production (Aspelund 2010). The journey usually starts with inspiration or identification. Two activities, inspiration and communication, occur and re-occur throughout the design process.

Inspiration

Creative effort drains energy from the designer. Inspiration is a source of energy for the creative process. "No matter what form it takes, inspiration is an energy that drives people to be creative and infuses their creations with life" (Aspelund 2010, p. 18). Note that Aspelund uses the term *life* here in the same sense as Alexander.

Inspiration has to be found. Designers should use their imagination to see the world as others have not. "The ability to create new worlds out of things that are all around us is indeed a valuable gift…. For creative endeavours, it is essential that designers tap into their imaginations, that they engage in playtime just like when they were kids. Playtime becomes the time designers use for experimenting and thinking" (Aspelund 2010, p. 22).

Nature, leading designs of the same product or different products, and the works of creative people (poets, musicians, etc.) are potential sources of inspiration. Engineering designers can increase the number and quality of inspirational resources by exploring nature, and studying the history of technology, acclaimed engineering designs, and creative works from the arts and humanities.

Identification

The aim of this activity is to create a design brief or specification and a project plan. The design brief defines the problem to be solved by the designer. There are no problems out there waiting to be discovered. Problems are artificial constructs created in our minds. Design briefs are generated through creative and systematic synthesis. A brief is an integrated description of the problem. One way to present an integrated brief is through use case scenarios.

Product design briefs must include the stakeholders, the context, product requirements and design constraints. The problem is not the designer's problem, but the problem for the people for whom the designer is designing. These are the designer's stakeholders. The brief must identify all the major stakeholders. The following people or groups should be considered (the list is not exhaustive):

- Customers (people who buy and use the product)
- Secondary users (people who are affected by the customers' use of the product)

- Manufacturer(s)
- Sellers/distributors
- Maintainers (people who will maintain the product)
- Disposers (people who dispose of the product when it is no longer required by the user)

Stakeholder requirements can be generated through interviews, quality function deployment (Cohen 1995), focus groups (Kreuger and Casey 2000) and idealised designing (Ackoff and Vergara 1981, Ackoff et al. 2006). The designer has a moral obligation to ensure that the stakeholders accept requirements that are legal and ethical. A more critical ethical issue is ensuring the world is left in a better state by the proposed design; that is, the design should be sustainable. Section II will discuss this aspect in more detail.

The brief should include the context for each stakeholder. An environmental specification (Chapter 2) will be part of the context, which will include factors that may affect the manufacture and use of the product, e.g., social and political factors. Section II presents a framework for exploring and defining the design context and associated risks.

Some other general points for briefs are:

- The design brief states what is known, but can include what is not known to provide a basis for future work (Cather et al. 2001).
- An adequacy assessment should be carried out to ensure the requirements are usable, feasible and valuable (Cagan 2008, Shigley et al. 2004).

Conceptualisation

This activity develops an initial design concept. The design synthesis techniques discussed previously can be used to generate ideas and form an initial solution. The solution is represented as a model. What constitutes a good concept? The simple answer is, it must have an unbeatable value proposition. It should be something that customers will say: "I must have this". Think of the iPhone and iPad. A good self-check here is to use FAB. FAB stands for features, advantages and benefits. What sells? Benefits and only benefits! A designer can talk about the features and advantages of the product as much as she or he likes, but stakeholders are only concerned with how the product improves their lives, what the benefits are to them.

The concept generation techniques described earlier will help you generate ideas. Polster et al. (2002) is a useful reference source for ideas.

Exploration/Refinement

A concept contains many hidden possibilities. The aim of this activity is to explore the concept and uncover the hidden possibilities. One or several will be selected to develop the design in more detail. It is important during this activity that the designer consider the design in context, how it will interact with the world.

The refinement of the design concept can be aided by studying engineering failures (Petroski 1994, 1996). Lessons learnt from engineering failures can help prevent future failures and improve the chance of success. NASA maintains a "learned lesson" database for this purpose.

The media a designer uses to express the concept "will influence the development of a concept; therefore, you need to be fluent with many different media" (Aspelund 2010, p. 97). Designers must be able to draw (freehand) and sketch. They should discuss their ideas with the stakeholders, and they should reflect quietly.

Definition/Modelling

This activity involves considering all the stakeholders' needs and ensuring the design meets them. A physical model is constructed to provide a realistic representation of the design so it can be evaluated from the stakeholders' points of view. The model can be a scale model and made of simple materials such as cardboard. The model is an exploratory tool to aid further conceptualisation and refinement of the design.

Communication

The aim of this activity is to develop the presentations needed to communicate effectively with stakeholders. The form of representation used by the designer to explore and evaluate the design may not be the best for presenting it to stakeholders. Furthermore, stakeholders may vary widely, and different modes of communication may be required to communicate effectively with them. Designers need to be conversant in different types of presentation media.

Production

Aspelund (2010) calls this production, but it is really a prototyping activity. A prototype is a physical model of part of the design or the whole design. The purpose of prototyping is to make a prototype that can be used to verify and perhaps validate certain aspects of the design. Design as a personal activity ends when the prototype is completed.

Design as a Phase

Design as a phase is a process that produces a model of an object or system that does not exist, but is capable of being brought into existence, and a critique of it by someone other than the designer(s). Design as a phase applies to products and manufacturing systems.

Design as a phase is a team activity. There must be at least one designer and one critic. On large projects there will be many designers and many critics. Furthermore, some of the sub-activities performed by a designer during design as a personal activity may be delegated to other people, freeing up time for the designer to focus on what designers do best: synthesis.

The creation of the design brief (identification sub-activity) is often taken to be a separate specification phase (Grady 2006, Shisko 1995) and delegated to specialists for execution. There is considerable controversy over whether this separation is a good idea or not. Aspelund (2010) and Brooks (2010) present strong arguments for including specification in the design phase, as part of the design process. First, the definition of the design problem and its solution are closely interlinked. Brooks (2010, p. 22) states that the hardest part of design is deciding what to design. In deciding what to design the designer must explore the problem at the same time as the design solution is being developed. Later he says that a design not only satisfies requirements, but also uncovers requirements. Second, the brief describes the general properties of the product being designed as well as the environment(s) within which it will exist. Creation of the brief is also designing (Chapter 16). Third, the separation often leads to the designer being excluded from the specification phase, resulting in products that do not meet customer expectations.

There are limitations to Brooks' and Aspelund's arguments. Brooks is discussing novel design in his book, innovative designs that change the world. He does not consider the more routine types of design where his arguments have less validity. Aspelund focuses on design as a personal activity and does not consider design teams. Notwithstanding these limitations, we believe identification should be part of the design phase for the reasons outlined above.

Concept integrity is easier for design as a personal activity compared to design as a phase because there is only one designer; the integration is carried out in one mind. It is a major issue for design teams. Concept integrity can be improved by making one or at most two chief designers responsible for the product architecture (Brooks 2010). Application of this rule depends on the number of levels of architecture for the product. One chief designer can design the architecture if there are a few levels. The other designers support the chief designer by designing parts of the product within the architectural constraints. The interfaces between the parts must be clean and clearly defined for this to work successfully. Several chief designers will be required if there are many levels of architecture. Each chief designer will be responsible for a few levels and is accountable to the next-higher-level chief designer (Chapter 16). The architectures at the different levels are integrated through their interfaces.

There is always one scarce resource for the design, and this must be budgeted for and tightly controlled (Brooks 2010). The scarce resource may be development cost (low-expense system architecture), unit cost (low-cost architecture), performance (high-performance architecture) or time to market (rapid development architecture) (Reinertsen 1997, Smith and Reinertsen 1998).

Design as a phase includes the seven sub-activities of design as a personal activity, critiquing activities and additional organisational and management activities to manage the design teams. Design team organisation is discussed in Chapter 16. Project planning and control methods are used to coordinate the activities of the design teams. System engineers are employed on large projects to coordinate activities across the interfaces, track the design scarce resource and carry out project planning and control and technical investigations for the designers.

Design requires reflection (Schon 1983, Thompson and Thompson 2008). It does not follow a step-by-step process. Instead, a designer has a reflective conversation with the design situation. The conversation guides the direction the design takes. Reflection is an acknowledgement of the tacit (unspoken) knowledge that designers have. Reflection is both a personal activity and a group activity (Thompson and Thompson 2008). There is further discussion on this topic in Chapter 17.

There are five main methodologies for designing product architectures—normative, rational, participative, heuristics and patterns, and six methods for critiquing designs—similarity, analysis, demonstration, simulation, examination and test (Chapter 2).

Design reviews are carried out during the design phase to assess whether the design has been completed sufficiently to proceed further. The reviews evaluate the design decisions and critiques. The type and number of reviews depend on the life cycle model used to control product development (Chapter 11).

Types of Design Projects

There are five major types of design projects (Ullman 2009): variation of an existing product (variational design), improvement of an existing product (extension design), development of a new design that will be made once (bespoke design), development of a new design that will be made in small quantities (batch design) and development of a new design that will be made in large quantities (mass design). The first two types of design projects are discussed in Chapter 12 and are not included here.

Bespoke design is the design of a product to meet the specific requirements of one customer. The full costs of product development are borne by the customer. Costs can be reduced by using standard, off-the-shelf parts and sub-systems, and configuring these to meet customer requirements. Existing manufacturing methods are used to make off-the-shelf parts and are hidden from the designer. Design effort is focused on configuration design. The designer will use existing processes and must know how to design the configuration to minimise construction and assembly costs. Bespoke design begins after a customer order is received.

Batch design is the design of a new product that will be made in small quantities for a few, but not many, different customers. In this situation it is not possible to meet all the different customer requirements due to the high cost. One or two leading customers are selected, and requirements are defined to meet their needs. Other customers benefit because the product will exceed their requirements. Some companies do not allow any variation in the product. Others allow some customer variation, with the variations being pre-designed. Customers select the variants from a catalogue.

Less off-the-shelf parts can be used compared to bespoke design, as the costs of newly designed parts can be spread over several products. Thus there is more flexibility to customise the design to customer needs. Existing manufacturing systems are normally used to make these products due to the low volumes. The designer needs to know what manufacturing systems will be used and how to design the products to minimise manufacturing costs. Product design and manufacturing architectures must be jointly designed if the design has a product family architecture.

Batch design starts before customers order the products. In most cases it is completed before customer orders are received. When development costs are high, companies will try to obtain customer orders and some form of pre-payment before the design is completed to improve cash flow.

Mass design is the design of a product that will be made in very large quantities for a large number of different customers. The high production volumes mean that dedicated manufacturing systems can be designed and constructed to make the products. Manufacturing is a major factor that must be considered during design. A common set of requirements is unlikely to satisfy all of the customers because there are so many; therefore companies allow for variation in the products. The manufacturing systems have to be designed to handle this variation. There is a very close coupling between product design and manufacturing system design. Mass design starts and is completed before customers order the products.

Design of Product Architectures

Chapter 4 discusses how small and medium-sized companies can compete globally. The successful companies export early and do not rely on home markets. They focus on niche markets, with a few products, and develop their core strengths. Batch design is the best approach under these conditions.

Batch design usually allows for some product variation and should provide a base for product extension. These conditions can be achieved with a product family architecture: platform and modular (Chapter 12). A product platform is the basis for developing product variants in the short term and extension in the long term. Platform extension involves technological upgrading of one or more sub-systems. The architecture for product variants and the extended platform is the same: same number and types of sub-systems and interfaces. The platform must be designed with an open architecture so that the variants and extensions can be created without much effort. This in turn requires clearly defined interfaces.

In a scale-based architecture, the platform extensions and variants are produced by changing specific design parameters. Platform architectural design involves defining the parameters that will not change, defining the scaling parameters, specifying the common sub-systems and their interfaces, and integrating the design with manufacturing. Mathematical techniques can be used to determine the appropriate design parameters for scaling, e.g., graph theory and matrices.

A modular architecture consists of modules. A common set of modules is contained in all products in the family. Other modules are substituted/added/subtracted to give customer variants. Thus a modular architecture consists of common and variant modules. The advantage of modular design is that modules can be made and assembled concurrently reducing the manufacturing lead time. The modules are designed for rapid assembly at the final stage.

A company that wants to implement any of the architectures above has to design both the product and the manufacturing system to meet the requirements of the new architecture. Furthermore, a new product concept may open up new ideas for other parts of the product life cycle, such as new distribution channels or new maintenance concepts. New product development should not be restricted to design of new products but to the design or re-design of other phases in the product life cycle. The other designs have to be jointly designed with the product. Where appropriate, DFX can be used to match designs to existing manufacturing processes and distribution systems. DFX means design for X, where X can stand for automation, assembly, packaging, injection moulding, etc.

Chapter 2 argued that products and systems should be designed to further progress to four ideals—plenty, truth, good and beauty. Progress toward truth and beauty can be achieved by designing new products to have life. The liveliness of products depends on product features that can be directly seen, heard, smelt and touched by users, i.e., the interface architecture. Design guidelines for designing interface architectures with life are (Alexander 2002, Chapter 5):

- Levels of scale. Products should be designed with centres that have a range of sizes. The sizes should be defined by well-marked levels or scale jumps. The scale jumps should be designed so the centre(s) at each level supports those at other levels.

- Strong centres. Centres should be strong visually. Centres should be designed to mutually support each other to create a field effect of centredness.

- Boundaries. Centres should be strengthened by boundaries. "Boundaries do the complex work of surrounding, enclosing, separating, and connecting in various geometric ways, but one vital feature is necessary in order to make the boundary work in *any* of these ways: the boundary needs to be of the same order of magnitude as the center which is being bounded" (Alexander 2002, p. 159).

- Alternating repetition. Centres intensify other centres by repeating. This is not simple repetition of one centre, but an alternating pattern of at least two centres.

- Positive space. The whole space should be filled by definitive positive spaces. There should not be odd spaces left over.

- Good shape. A good shape is a shape that consists of centres that cohere as a whole. Shapes should not be amorphous. Alexander (2002, p. 183) provides a partial list of properties that make up a good shape: high degree of internal symmetries, bilateral symmetry (almost always), a well-marked centre, the space it creates next to it is also positive, it is strongly distinct from what surrounds it, it is relatively compact, and it has closure.

- Local symmetries. Centres are reinforced by local symmetries. This is different from overall symmetry. Generally, an overall symmetry does not promote life.

- Deep interlock and ambiguity. Centres should be designed to interlock and overlap with their surroundings. It should be difficult to visually separate the centre from the surrounding.

- Contrast. Create designs with contrast. Typical forms of contrast are black-white or dark-light, solid-void, busy-silent, red-green, etc. (Alexander 2002, p. 200).

- Gradients. Gradients create softness. Gradients depend on the properties being adjusted to, e.g., light. Gradients are produced by varying centres in size, spacing and intensity. Use product gradients to follow natural gradients, to provide a transition from one scale to another and to soften the look of products.

- Roughness. Design products to have a certain amount of roughness, a certain amount of irregularity. Repeat patterns should not be exact.
- Echoes. An echo refers to a similarity (family resemblance) between a set of centres. The similarity is based on angles and families of angles.
- Voids. Voids are empty spaces. They are intense centres. All centres should be designed to have some emptiness.
- Simplicity and inner calm. Design products to have simple shapes. More complex shapes can be created out of these simple shapes.
- Not separateness. The product should be designed to visually fit into its surroundings, to be a part of the surroundings and not separate from them.

Design Standardisation

Design standardisation is a set of rules limiting the number of design options for parts: individual components, sub-assemblies and modules. Standardisation reduces product life cycle costs and improves product reliability and durability. Life cycle cost savings occur because:

- Designers do not have to design the re-used parts.
- Critics do not have to validate and verify the parts.
- The lower variety results in higher production volumes, reducing manufacturing costs.
- A smaller variety of parts reduces inventory and obsolescence costs.
- Standard parts reduce assembly time.
- Standard parts reduce maintenance/service downtime and spare part costs.

Reliability and durability are improved because parts are used, and therefore tested, in a wider variety of conditions than would be the case if they were used solely for one product.

There are national and international standards for a wide variety of parts, such as nuts, bolts, washers, etc. These standards should be used. A list of approved standards should be maintained and made accessible to designers. There may be cost advantages in restricting the range of standard parts that can be used. A list of the preferred sizes should be maintained and made accessible to designers. Both lists can be easily implemented if a computer-aided design (CAD) system is used for design. Approved parts can be included in the CAD system part libraries, and selection rules can restrict the parts that designers can select.

Standardisation at a higher level is more complicated. It depends on whether standardisation is for spares or to meet customer requirements. First, consider spare parts. Spare parts are provided for products that wear out or fail over time. Spare parts are not used for products that cannot be separated, e.g., a welded frame, or for disposable products. In these cases, the complete products are replaced when damage or failure occurs.

The provision of spare parts depends on the maintenance concept for the product. Often it is better to exchange complete sub-assemblies during maintenance and carry out repairs offline, as this reduces the product downtime. The repaired sub-assemblies are placed in stock for future use. The stocking of spare parts at the sub-assembly or module level is also used to reduce the variety of parts that have to be stocked, lowering inventory costs. This does not mean that other parts will not be provided to customers should they need them. In these instances, each customer request will be evaluated and a separate quotation given for cost and delivery time.

The provision of spare parts must be considered early in the design process, as it can have a considerable impact on how the product is designed. The maintenance concept and design concept should be developed concurrently (Appendix C).

A product family architecture is the most effective way of achieving standardisation for customer requirements. Common sub-assemblies and modules are used for different products. The determination of these depends on the type of product architecture that is selected and how it is designed (Chapter 12).

Design and Manufacture

Companies use five strategies to design and make products for customers. Usually companies will employ only one strategy, but it is possible to mix strategies. Hydroline Oy, a Finnish company that designs and manufactures hydraulic cylinders, uses a mix of three strategies: design-to-order, make-to-order and deliver-to-order. Design-to-order is used for customised cylinders for low volumes, make-to-order is used for customised cylinders for high volumes, and deliver-to-order is used for standard cylinders.

The first approach is *engineer-to-order* (Figure 3.3). Engineer-to-order is applicable when the technology and product design are different for each customer. Technology development *and* design are included in the order life

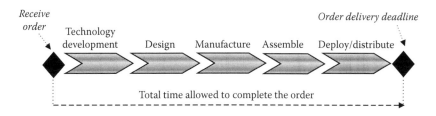

FIGURE 3.3
Engineer-to-order cycle.

cycle and must be completed within the promised customer deadlines. In other words, new technology is developed through product development.

Engineer-to-order incurs the highest technical risk because technology development and design are carried out under customer order deadlines. The outcome of technological development is very uncertain, and there is always the risk that technical requirements may not be achieved in time or at all. Risk can be reduced by running alternative development projects in parallel. One project will focus on stretching the frontiers of knowledge and is high risk. The other project will use mature technology and adapt it to the new situation and is low risk. The low-risk project is a backup in case the high-risk project fails.

Design risk is less uncertain than new technology development. However, it is high in this case, as the design will be based on new technology that has not been proven. Design critique is critical for success. Computer-based analysis and simulation should be used to rapidly iterate the design-critique cycle. Accelerated stress testing should be used to find design mistakes and improve reliability and durability. It is the fastest way to find design mistakes, but comes at a price: increased development costs. Manufacturing risk may be high if the design requires new manufacturing processes.

The strategy is best suited for one-off projects where radical innovation is desired. The design phase should be executed using bespoke design.

Technology development is carried out separately and independently of design in the *design-to-order* strategy (Figure 3.4). Design does not proceed until the technology has been proven ready for design. Design is included in the order life cycle. This type of product development incurs medium technical risk—technology risk is avoided, but design risk is there because design is carried out under customer order deadlines. Design critique is critical for success, and the comments above apply here also.

Design risk can be minimised by using mature technologies, design techniques and processes. Novelty can be achieved through innovative configurations or applications of the product.

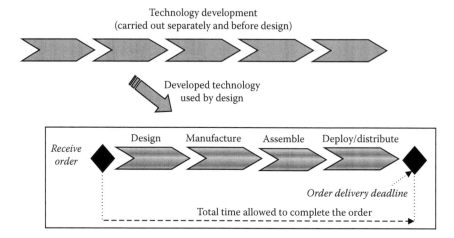

FIGURE 3.4
Design-to-order cycle.

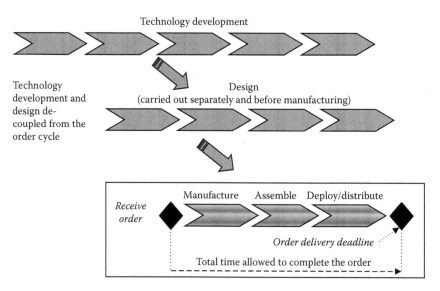

FIGURE 3.5
Make-to-order cycle.

The strategy is suitable for one-off and small-volume production. The design phase should be executed using bespoke or batch design.

In *make-to-order* (Figure 3.5), technology development and design are carried out separately before the order cycle. The designs are standard for all customers, including pre-designed customer variants, although a small amount of customisation can be permitted if required. This type of development

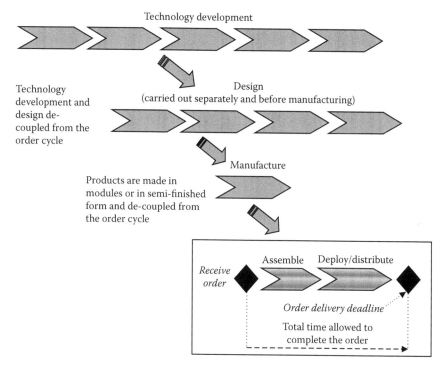

FIGURE 3.6
Assemble-to-order cycle.

incurs the lowest technical risk because technology development and design are not executed under order deadlines.

The strategy is suitable for batch and mass production. The design phase can be executed using batch and mass design.

Assemble-to-order (Figure 3.6) is similar to make-to-order. The key difference is that semi-finished products are made and stored before customer orders are received. Products are assembled and delivered after an order is received. The strategy is used when customers demand very short lead times between ordering and delivery.

Deliver-to-order (Figure 3.7) is a make-for-stock cycle. Finished products are made and assembled and stored in inventory. The products are delivered on order. The strategy is used for standardised products that are sold to a range of customers; hence overall demand is relatively constant. Supply-chain inventories should be vendor managed. A vendor-managed inventory (VMI) is a supplier inventory located at a customer site where the supplier is paid only when products are withdrawn from the inventory. The supplier must maintain the inventory to meet customer demands, but has to bear the inventory carrying costs and be able to replenish it on short notice.

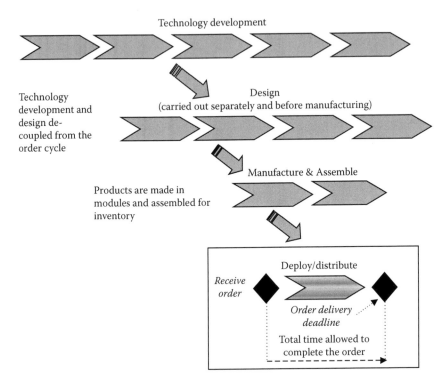

FIGURE 3.7
Deliver-to-order cycle.

Exercises and Problems

1. Briefly describe design as a personal activity.
2. Briefly describe design as a phase.
3. Briefly describe the design and manufacture cycles and give an example of each.

References

Ackoff, R.L., Magidson, J., and Addison, H.J. 2006. *Idealized design*. Englewood Cliffs, NJ: Prentice Hall.

Ackoff, R.L., and Vergara, E. 1981. Creativity in problem solving and planning: A review. *European Journal of Operational Research*, 7, 1–13.

Alexander, C. 2002. *The nature of order. Book 1: The phenomenon of life*. Berkeley, CA: Center for Environmental Structure.

Arieti, S. 1976. *Creativity: The magic synthesis.* New York: Basic Books.

Aspelund, K. 2010. *The design process.* 2nd ed. New York: Fairchild.

Brooks, F.P., Jr. 2010. *The design of design.* Upper Saddle River, NJ: Addison-Wesley.

Buzan, T., and Buzan, B. 1996. *The mindmap book.* New York: Plume.

Cagan, M. 2008. *Inspired: How to create products customers love.* Sunnyvale, CA: SVPG Press.

Cather, H., Morris, R., Philip, M., and Rose, C. 2001. *Design engineering.* Oxford: Butterworth-Heinemann.

Chakrabarti, A. (ed.). 2002. *Engineering design synthesis.* London: Springer.

Cohen, L. 1995. *Quality function deployment.* Reading, MA: Addison-Wesley.

Csikszentmihalyi, M. 1997. *Creativity.* New York: Harper Perennial.

De Bono, E. 1983. *Lateral thinking: A textbook of creativity.* London: Penguin Books.

Eberle, B. 2008. *SCAMPER.* Waco, TX: Prufrock Press.

Gordon, W.J.J. 1961. *Synectics.* New York: Harper & Row.

Grady, J.O. 2006. *System requirements analysis.* Burlington, VT: Academic Press.

Hubka, V., and Eder, W.E. 1988. *Theory of technical systems: A total concept for engineering design.* London: Springer-Verlag.

Kreuger, R.A., and Casey, M.A. 2000. *Focus groups.* 3rd ed. Thousand Oaks, CA: Sage Publications.

Novak, J.D. 2010. *Learning, creating, and using knowledge.* 2nd ed. New York: Routledge.

Osborn, A.F. 1963. *Applied imagination.* New York: Charles Scribner's Sons.

Pahl, G., Beitz, W., Feldhusen, J., and Grote, K.-H. 2007. *Engineering design: A systematic approach.* 3rd ed. London: Springer-Verlag.

Petroski, H. 1994. *Design paradigms.* Cambridge: Cambridge University Press.

Petroski, H. 1996. *Invention by design.* Cambridge, MA: Harvard University Press.

Polster, B., Neumann, C., Schuler, M., and Leven, F. 2002. *The A–Z of modern design.* London: Merrell.

Prince, G.M. 1970. *The practice of creativity.* New York: Harper & Row.

Pugh, S. 1991. *Total design.* Wokingham, UK: Addison-Wesley.

Rantanen, K., and Domb, E. 2007. *Simplified TRIZ.* 2nd ed. Boca Raton, FL: Taylor & Francis.

Reinertsen, D.G. 1997. *Managing the design factory.* New York: Free Press.

Schon, D. 1983. *The reflective practitioner.* New York: Basic Books.

Shigley, J.E., Mischke, C.R., and Budynas, R.G. 2004. *Mechanical engineering design.* 7th ed. New York: McGraw-Hill.

Shisko, R. 1995. *NASA systems engineering handbook.* Washington, DC: NASA.

Smith, P.G., and Reinertsen, D.G. 1998. *Developing products in half the time.* New York: John Wiley.

Suh, N.P. 1990. *The principles of design.* Oxford: Oxford University Press.

Thompson, S., and Thompson, N. 2008. *The critically reflective practitioner.* Basingstoke, UK: Palgrave Macmillan.

Ullman, D.G. 2009. *The mechanical design process.* New York: McGraw-Hill.

Zwicky, F. 1969. *Discovery, invention, research through the morphological approach.* Toronto: Macmillan.

4

Manufacturing System Design

Introduction

There are five main types of manufacturing: project, jobbing, batch, mass and process (Figure 4.1). Process manufacturing is not discussed in this book. It is very high-volume, continuous flow production; examples are oil refineries and chemical and pharmaceutical manufacture. Very high-volume production of discrete parts sometimes approaches near continuous production and can be classified in this group also, e.g., semi-conductor foundries. The other four types apply to discrete part manufacture.

Project manufacturing is the manufacture of one-off products. Usually the products are very large and complex, e.g., a process plant, a power plant, a factory or a communications satellite. The appropriate product architectures for this type of manufacture are customised and modular. The manufacturing equipment is general purpose and arranged in fixed position layout or according to the work breakdown structure.

Jobbing manufacturing involves making a wide variety of different products for a wide variety of different customers. The volume of each component or product is low, from a few items up to 50 per year. The appropriate product architectures are customised and modular. General purpose equipment is used and usually arranged in a functional layout. Typical examples are engineering model shops, general purpose machine shops and small foundries.

Mass manufacturing is high-volume production of one or a few different products; typically production volumes exceed 10,000 per year, but can go as high as millions per year. Product architectures should be optimised. Product family platform architectures are also very effective, as Volkswagen (Wilhelm 1997) has clearly demonstrated. Equipment is specialised and arranged in a line flow. Typical examples are automotive, computer, mobile phone and household appliance manufacturers.

Batch manufacturing lies in between jobbing and mass manufacture. Typically batch manufacture involves production volumes from 100 to 10,000 per year depending on the industry. The most appropriate product architectures are modular and product family platform. General and special purpose equipment is used and the best arrangement is a cellular layout. Typical

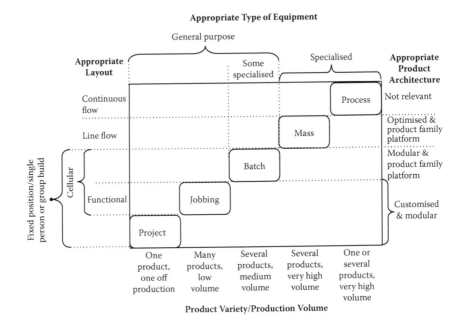

FIGURE 4.1

Types of manufacture. (Adapted from Hayes, R.H., and Wheelwright, S.C., *Harvard Business Review*, 57(1), 133–140, 1979; Slack, N., et al., *Operations Management*, Harlow, UK: Pearson Education, 2004; Miltenburg, J., *Manufacturing Strategy: How to Formulate and Implement a Winning Plan*, Portland, OR: Productivity Press, 2005.)

examples are industrial electronics manufacturers, industrial equipment manufacturers and niche market consumer product manufacturers.

We shall focus on batch manufacturing, as this is the most appropriate type of manufacture for small to medium manufacturing companies (SMMCs). Figure 4.2 shows a concept map for the chapter. The chapter explains how to design a factory and supply chain based on part manufacturing and assembly cells. We talk about group technology (GT), which is a methodology for designing cells to make parts, and how it can be applied through a technique known as production flow analysis. We present a design procedure and guidelines for assembly and test cells. The chapter concludes with design guidelines for global manufacturing.

Project Manufacturing

Project manufacturing is the manufacture of one unique product. Large, complex systems are the most difficult to make and construct and are undertaken by large companies as the prime contractor. Small companies

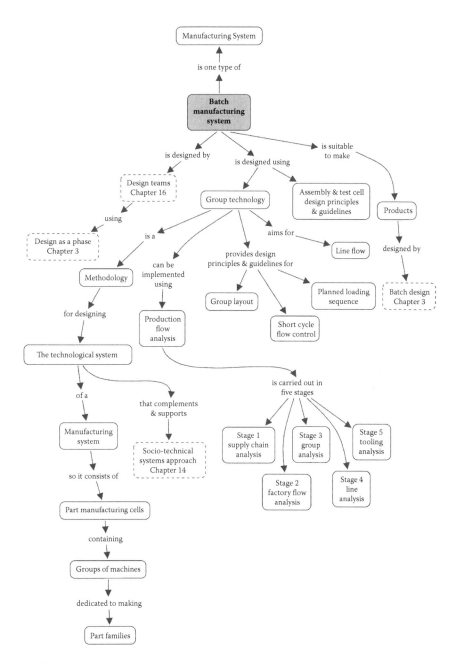

FIGURE 4.2
Concept map of Chapter 4.

manufacture components and small assemblies under sub-contract from the prime contractor.

There is considerable market uncertainty for project manufacturers. They have to bid for projects. The project environment is highly competitive. Small companies may submit as many as 20–30 bids for different projects before they win a contract. The market uncertainty in this environment is:

- Timing of customer orders (winning a bid)
- What is to be made (scope of work)
- How the component or product is to be made
- How much work is required

A bid contains the following items:

- Scope of work (what is to be done). The scope should be defined in terms of clearly defined deliverables. The scope may include how the product will be made.
- The price. An estimate has to be made of how much work is required to calculate this.
- Delivery promise. The dates for when the customer will receive the deliverables.
- Quality standard. The quality standard(s) that will be used to design, make, deliver and install the product.

Strategic planning for this type of company involves bidding for and winning contracts. A bid has a major impact on the execution of the project. Poor bid preparation greatly increases the risk of project failure. Poor estimation of the scope of work can lead to staff not being able to perform the project tasks to the required quality standards, or the required technology not being available when needed, or under-estimation of the amount of work. The end result is inaccurate pricing and delivery promises. Risk assessment is also critical. Poor risk estimation can lead to bankruptcy and project cost and schedule overruns.

The appropriate project planning and control systems at the operational level are critical path method (CPM) and program evaluation and review technique (PERT). A scope change control system manages requirements and scope changes.

Project manufacturers usually employ a small core group of full-time staff grouped functionally into departments. They are assigned to projects on a part-time or full-time basis. Additional temporary staff are employed to meet the needs of specific projects. Quality is assured by employing highly skilled staff and rigorous project quality standards and procedures. Most equipment is general purpose, but specialised equipment may be justified for large projects. Equipment is arranged in a fixed position layout or may be matched to the project work breakdown structure.

Jobbing Manufacturing

Jobbing manufacturers are typically small owner-operated companies. By small is meant a staff size of 50 employees or fewer. The market environment is highly competitive and closely corresponds to the economic model of perfect competition. In this type of market there are many companies, each supplying only a small proportion of the total demand. Companies can enter or leave the market freely. There are also many customers, each consuming only a small proportion of the total demand. Each company produces to the same quality specification from a customer's point of view. The customers have no preference for specific products, or for specific manufacturers, as long as their requirements are met.

A given company has plenty of customers that it meets at random: either they come to it or it searches for them, or both. Whenever a company and a customer meet there is a certain probability of making a sale. This depends on whether the price is acceptable to the potential customer. The best planning approach is to make a sale when the chance arises and attempt to maximise the profit on each sale (tactical planning).

The market uncertainty is very high, as the scope of work is not known until a customer order is received; it is the same as for project manufacturers. Due to the high level of market uncertainty the appropriate strategy for a company is to use general purpose equipment and skilled staff, who can provide a range of services. The quality of output depends mainly on the skilled staff. There may be some standardisation of work.

Tight cash flow control is essential to ensure a company has sufficient liquidity between orders, as customers pay on completion of the work. Progress payments can be used to ease liquidity. The appropriate production planning and control systems are job loading, project planning and control using PERT or critical path method (CPM), and period batch control.

The organisational structure is normally a functional structure. The equipment is usually arranged in a functional layout, although group technology layout can be used by companies operating in niche markets.

Batch Manufacturing

Batch manufacturers are small to medium-sized companies, ranging in size from 50 to 1,000 employees. The market environment corresponds to the economic model of imperfect competition. The batch manufacturer has built up brand and company loyalty with a cluster of customers. A customer will not change brands unless the brand is not available or a price differential exists for a relatively long period of time between the brand the customer prefers and other brands available in the market.

These companies use standard designs for their products, and hence reduce market uncertainty in terms of what is to be done and how it is to be done. Given its cluster of customers, a company can learn their habits and thus more closely match its products and services to their needs. This raises the entry barrier for new competitors trying to enter the same business.

The only uncertainties are how much work (size of order) and timing of orders. These can be estimated from past sales and knowledge of the customer characteristics of the cluster. This allows the manufacturer to plan production rates and inventory levels to provide good service at a reasonable price. Hence the most important planning approach is strategic planning, supported by tactical planning.

The use of standardised designs means special purpose equipment can be used to reduce manufacturing costs. However, there will be some variability in customer requirements, and so general purpose equipment may be required. Skilled staff are generally employed to set up equipment and for engineering design. Semi-skilled staff are employed to operate the equipment, and unskilled personnel are employed for administrative work. Quality control is achieved through skilled staff and the special purpose equipment. There will be some standardisation of work.

Timing and cost are very important for winning an order. Work is carried out in batches to reduce costs. The logical complexity of allocating batches to people and equipment means that production and inventory control become very important. Risk, and cost, can be reduced considerably by shortening throughput times. Cash flow regulation is important, and so is credit control. The best situation is long-term credit from suppliers and short-term credit for customers. The worst situation is exactly the reverse. The appropriate production planning and control systems are period batch control, standard batch control, lot-for-lot materials requirements planning (MRP) and just-in-time (JIT) production.

The organisational structure should be based on group technology. A line/staff (operation/support) form is appropriate. Group layout is the best for part manufacturing. The part manufacturing cells are supported by assembly and test cells.

Mass Manufacturing

Mass manufacturers are large companies producing thousands, even millions, of products per year. The market environment corresponds to the economic model of oligopoly. An oligopoly is a market of finite size comprised of a large number of small customers and a small number of large manufacturing companies, each of which is competing for a larger share of the market. Each company has the same range of products, but quality, style, service and prices vary from company to company due to different management, design and development, manufacturing and marketing capabilities.

New companies are generally excluded from the market for two reasons. First, local companies are excluded because the entry costs are too high. Second, large foreign companies cannot compete because often an existing oligopoly will persuade consumers and the government that foreign competition must be restricted or excluded. Generally this takes the form of imposing high duties or quotas on foreign products.

A company wanting to increase market share can do so by (1) reducing costs, (2) producing better quality products with more attractive styling and better functionality, (3) providing a better service, or (4) some combination of these. In this environment a company can affect both the probability of making a sale when it meets a customer and the probability of meeting customers. Advertising is a powerful tool for increasing these probabilities.

The key planning approach is operational planning, supported by strategic and tactical planning. Each company has to accept that what it knows is also known to the other competing companies, and that the cluster of customers that it is trying to capture is also desired by other competing companies. Consequently, each company has to try to outsmart the others, and they will do likewise; this generates the market dynamics. Thus not only does a company have to plan where it wants to go and how to get there, but it must also plan how to deal with its competitors. Two strategies are commonly used: absorption and parasitism. Absorption is a means of gaining market power by taking over competitors or through backward or forward integration—increasing the length of the supply chain that is directly controlled. Parasitism is used by small companies to gain power by linking up with a mass manufacturer as a sub-contractor or supplier.

Sometimes market stability is achieved by members of the oligopoly making agreements with each other, e.g., forming market share agreements, price fixing agreements or agreements not to poach personnel. These may achieve stability for the companies concerned, but customers often lose out. Many governments legislate against these kinds of agreements to protect consumers.

The very large market size means that companies can afford to use special purpose equipment. The main costs are now associated with the plant and equipment and cost control focuses on control of inventories. Marketing becomes vitally important in this environment because products cannot be stored for long without incurring high inventory costs. The production equipment must not remain idle either because of its high capital cost. Hence the market must be stimulated frequently to demand the product at a rate that is economical to produce.

Control of quality is largely built into the plant and equipment. A significant labour component occurs at some stages, particularly assembly, and hence worker control of quality is also important. Maintenance of plant and equipment becomes a major engineering activity. A large number of unskilled staff are employed to operate the equipment. Work can be standardised to a very high degree. As a result, a lot of personnel are employed in staff functions for the sole purpose of standardising work.

Quantity and timing control is largely built into the plant and equipment. Costs can be reduced by reducing setup times so that products can be made in smaller batches and by smoothing the production program. This allows the production system to more closely track market demand and increases the inventory turnover.

The stability of the production processes allows the use of optimisation techniques to improve the processes over time, e.g., integer, linear and non-linear programming, dynamic programming, and queuing theory. The techniques are also used to optimise resource allocation and for determining optimal product mix. The appropriate production planning and control system is JIT.

The most common organisational structure is a line/staff (operation/support) form in the factories. The equipment is arranged in dedicated manufacturing lines or cells. Flexibility can be provided by designing the lines to make several products and running the lines with small batch sizes. Mass manufacturers are usually divided into divisions at the corporate level.

Part Manufacturing Cells: Group Technology

Line flow is the most efficient means for arranging the flow of materials through a factory. In this type of flow, the manufacturing processes and machines are arranged close together in the sequence in which they will be used. Line flow is common in mass manufacture, where whole factories will be arranged in flow lines to manufacture and assemble one product. The large volume of products makes this economically viable. The advantages of line flow are low work-in-progress inventories and short lead times.

The small volume of products, frequent changes in products and high variability in demand encountered in jobbing and batch manufacture mean it is not economically viable to have single-product flow lines. However, it is possible to divide the parts of different products into families and the equipment into groups so that one group of machines, a group technology cell, makes one family of parts. The machines within a cell can be arranged in a flow line and the cells can be sequenced in a flow line. The higher volume of families of parts makes this economically viable.

Group technology is a methodology for designing, implementing and operating cells based on families of parts. It applies to part manufacture but not assembly. Group technology is really quite old. Some companies were practicing group production methods before group technology was developed as a formal methodology; Gallagher and Knight (1973) discuss an example as early as 1925. Group technology was popular in the 1950s through to the 1970s, but has been largely overlooked since then. This is unfortunate, as the reasons for implementing it are as applicable today as they were in the past, despite improvements in machining technology and manufacturing equipment.

Group technology originated from the work of Professor Mitrofanov of Leningrad University on improving the efficiency of using lathes (Burbidge 1979). He showed that loading similar parts on lathes one after the other reduced setting times, increased capacity and reduced tooling costs. Application of these results in industry gave rise to the single-machine approach. Changes in loading to single lathes were implemented, and that is all.

The next development was to arrange supporting machines next to the lathes to create groups with different machines. Each group made a family of parts. The part families were determined by eye. This stage of group technology is sometimes referred as the pilot group stage.

The next stage of development was to use formal methods to divide complete workshops into groups. Initial work focused on classification systems: design classification first, then production classification. The developments clearly demonstrated that group technology could be applied to whole factories with beneficial results, and that it required major changes in organisation, production and control systems, payment systems, etc. It was during this latter stage that Burbidge (1971, 1975) proposed production flow analysis (PFA) as a method for finding part families and dividing equipment and workers into groups. PFA is applicable to supply chains, factories and departments in factories.

The key features of group technology (Burbidge 1975) are:

- Group layout: Laying out the set of machines/equipment in each group closely together and separated from other groups of machines/ equipment.
- Short cycle flow control for ordering of parts/components to be made by each group.
- A planned loading sequence.

The advantages of group technology are due to reduced setting times, group layout and short cycle flow control. Although each of these features will produce benefits if implemented separately, together they reinforce each other to produce major savings in costs.

The advantages of setting time reduction are:

"1. An increase in machine capacity
2. A reduction in the tooling investment
3. A reduction in setting costs
4. A reduction in operation costs" (Burbidge 1975, p. 38)

The advantages of group layout are:

"1. Reduced throughput time
2. Improved ability to follow market changes
3. Reduced stocks
4. Centralisation of responsibility
5. Reduced handling and setting costs
6. Simplification of paperwork
7. Reduced indirect labour
8. Improved human relations
9. Reduced investment per unit output" (Burbidge 1975, p. 40)

The advantages of short cycle flow control are:

"1. Reduced materials obsolescence
2. Reduced direct material cost
3. A contribution to the following savings already described under group layout:
    (i)   Reduced throughput time
    (ii)   Improved ability to follow market changes
    (iii) Reduced stocks
    (iv)   Simplification of paperwork
    (v)   Reduced indirect labour
4. The elimination of inter-process stores" (Burbidge 1975, p. 46)

The reduction in setting times increases production capacity. The reduction in throughput time, shorter delivery times and greater flexibility to follow market requirements all tend to increase sales output. At the same time, significant cost savings are achieved: reduction in material costs, handling costs, production control costs, etc. The reduction in tooling investment and the elimination of inter-process inventories reduce the investment needed to sustain a given level of output.

The only limitation on group technology is that it should not be used if line flow is economically feasible. There may also be an issue convincing management to implement group technology in a company. Machine utilisation will be lower, in general, with group technology than with a functional layout. Standard accounting systems measure costs associated with resource usage (machine utilisation), but not the value of lead time. Therefore a change to group technology will often show as an increase in cost. In order to properly justify group technology it is essential that the value of lead time be estimated. The costs savings in lead time reduction can be used to offset possible increases resulting from lower equipment utilisation. Lead time measurement is discussed in Reinertsen (1997) and Smith and Reinertsen (1998).

There are three main methods for dividing parts into families and machines into groups:

- Design classification: Parts that are similar in form, geometry and design function are classified into a family that can be made in the same cells. The limitation of this method is it does not find parts that are dissimilar in form but can be made in the same cell. For an extensive discussion on design classification systems see Gallagher and Knight (1973) and Arn (1975).
- Production classification: Parts that have similar production features, such as materials and tolerances, are classified into families and these are used to form the cells. This method finds dissimilar parts with similar production methods. For an extensive discussion on production classification systems see Gallagher and Knight (1973) and Arn (1975).
- Production flow analysis (Burbidge 1971, 1989): The production paths (routes) of parts are analysed to find parts with common paths. Families are based on common paths and the machines in the paths form the groups.

Production flow analysis (PFA) will be described here, as it is the simplest and most general method for forming groups. The major aim of PFA is to design the simplest material flow system by eliminating flow paths with low flow rates. PFA is carried out in five stages:

1. Supply chain analysis
2. Factory flow analysis
3. Group analysis
4. Line analysis
5. Tooling analysis

Supply chain (network) analysis is used to structure the supply chain in the most efficient manner. Network analyses are used to model the flow of products and materials along the supply chain. Paths with low flow rates are eliminated. This is achieved by eliminating or re-designing products on low flow rate paths. Alternatively, low flow rate paths can be eliminated by re-allocating equipment between factories.

Factory flow analysis aims at developing a simple and efficient inter-departmental flow system within the factory. Each department will contain one or more work cells. Network analyses are used to model the flows of components and materials, and paths with low flow rates are eliminated. This is achieved by moving equipment from one department to another or by re-designing the parts on low flow rate paths so they flow along the main paths.

Group analysis considers each department and divides their parts into families and the equipment into group technology cells. The aim is to have minimal interaction (maximal independence) between the groups. The input to this stage is the manufacturing requirements for the parts, which are specified in the form of flow paths or process routes. The information is represented in a part-machine matrix, with the parts as columns and the machines as rows. Grouping consists of partitioning this matrix to form independent groups. Partitioning determines the part families and machine groups at the same time. A number of mathematical algorithms have been proposed to do this (Singh and Rajamani 1996):

- Direct methods: Bond energy algorithm, rank order clustering algorithms, direct clustering algorithm and cluster identification algorithms.
- Similarity coefficient clustering methods: Single-link clustering algorithms, average link clustering algorithms and linear cell clustering algorithms.
- Mathematical programming and graph theoretic methods: P-median model, assignment model, quadratic programming model, graph theoretic models, etc.

For simple cases, partitioning can be carried out manually, by trial and error, using spreadsheets.

The analysis may find exceptions, parts that do not fit into any of the families. These should be eliminated or alternative processing methods found.

Line analysis simplifies the flow of materials between machines inside each group. The aim is to lay out the equipment in a sequence, without any reverse flow. Parts can skip machines in the sequence, but there should be no backtracking. Network techniques are used to determine the line layouts.

Tooling analysis is a study of the tooling requirements for each machine in a group technology cell. The aim is to reduce setup times through rapid or multi-tool setups. Multi-tool setups are defined by finding families of parts that can be made with the same set of tools (a tooling family). Matrix methods are used to find the tooling families.

Short cycle flow control is an essential component of group technology. The parts made by group technology are assembled into products. The number of parts in each product is known and fixed and should be ordered in matched sets, i.e., in the exact quantities required for assembly. There are four ordering systems that make parts in matched sets and give short cycle flows.

Period batch control: The year is divided into a number of equal periods, e.g., week or 2-week periods. Parts and sub-assemblies for the products are ordered in matched sets. The number of assembled products to be completed each period varies from period to period to meet customer demand. In period batch control, as described by Burbidge, the manufacturing process is divided into stages with throughput times that match the planning period. This is to reduce the amount of work involved for planning and control. For example, consider a manufacturing process in a factory that consists of four stages and a 1-week planning period. The total throughput time for the process will be 4 weeks. In any week there will be four cycles in process in the factory. The number of cycles in process can be reduced by reducing the number of stages.

Standard batch control: The number of assembled products to be completed is the same for each planning cycle, but the planning time interval varies to match demand. Parts and sub-assemblies for the products are ordered in matched sets. The lead time offset is the same for each order because the production quantity is the same. Note that for this type of ordering it is not necessary to divide the manufacturing process into stages with the same lead times.

Lot-for-lot MRP: Materials requirements planning is a program-controlled, computer-driven system for planning and controlling manufacturing materials and parts. Demand is estimated, and based on this the number of parts to be made is planned. It is a push system. Orders are released at the beginning of the process and parts are pushed through to meet deadlines. MRP systems can be programmed to use any ordering rule. Most of the ordering rules *do not* make parts in matched sets. However, MRP systems can be set up to use period batch or standard batch ordering; this is referred to as lot-for-lot MRP.

Just-in-time (JIT) production: JIT is a base stock control system. In-process inventories are located between machines or processes (the base stock inventories). Each time a part is withdrawn from one inventory an order is sent to the preceding process to replace it. It is a demand-based pull system. An order is released at the end of the process (the end of the assembly line), triggering off orders in the preceding processes and pulling parts through the system.

Cells making cheap common parts and spare parts can use cheaper ordering systems than those described above, e.g., *Q-ordering, P-ordering,* or *maxmin* ordering. Cells making long lead time and expensive parts can have their own separate ordering cycles using *component batch ordering.*

The following design guidelines apply to group technology cells (Burbidge 1971, 1975, 1979, 1989).

Between Cells

- Process routes. Simplify the process paths. Eliminate paths with low flow rates.
- Group definition. Ensure there is minimal interaction (maximum independence) when defining groups.
- Part family definition. Eliminate exceptional parts that do not fit into any family, or else develop alternative production processes to make these parts so they fit into one of the current families.
- Task protection. Ensure that once materials enter a group they can be processed wholly within that group without external disruption.
- Boundary regulation. Ensure that once materials leave a group they do not return to the group.
- Layout between groups. Groups should be sequenced and located according to the material flow.
- Boundary definition. The set of machines/equipment in each group should be located closely together and separated from other groups of machines/equipment (group layout or just-in time layout). This protects the primary task and gives clear ownership of the equipment to the work group.
- Production planning and control: Main ordering system. Use short cycle flow control for ordering of components to be made by each group (period batch control, standard batch control, lot-for-lot MRP, and JIT).
- Production planning and control: Cheap common parts. Use a cheap and simple ordering system such as Q-ordering, P-ordering or maxmin ordering.
- Production planning and control: Long lead time and very expensive parts. Use a separate ordering cycle, component batch ordering.
- Costing system. Each group should be set up as a separate cost, performance or profit centre.

- Organisation. Each group will be independent with its own members and have its own group leader. The groups are accountable for coordination and control of their activities within the group.
- Worker flexibility. Workers in a group should be multi-skilled to improve flexibility in work assignments.

Within Cells

- Process path. Minimise reverse flow. The ideal is no reverse flow.
- Layout in the group. Arrange the equipment/machines in a line flow.
- Tool storage. Special tools required by the group should be stored inside the group.
- Machine setups. Use rapid or multi-tool setups.
- Housekeeping. Use 5S (sort, set in order, shine, standardise, sustain) to design each workspace in the cell.
- Production planning and control. Use a planned loading sequence for making parts in the cell.

Assembly and Test Cells

This section discusses how assembly and test cells should be designed. It does not include part assemblies. A part assembly is an assembly where a number of parts are attached to each other in a fixed arrangement. The assembly is then treated as an individual part and undergoes further material processing. An example is a bicycle frame. It is composed of a number of tubes (parts) that are welded together to form an assembly, the frame. The frame is further processed by machining and grinding, followed by painting. Part assemblies can be made using part manufacturing cells. The design guidelines in the previous section apply.

The quality of an assembly depends on the quality of the parts being assembled and on the integration. Inspection and testing is required to verify the quality of integration. Unfortunately, the literature on lean manufacturing gives the impression that no inspection or testing is necessary because quality is "built in to" the product. This is not true. Some inspection and testing is required. Inspection and testing may be carried out in the assembly cell at an inspect/test station or in a separate cell (e.g., testing of aero-engines). In both cases, the station or cell should be located to follow the sequence of the workflow.

An assembly line is a common method for assembling products. An assembly line is not necessarily an assembly cell. Take, for instance, the traditional, long, machine-paced assembly line. This does not constitute an assembly cell. The workers have no control over the pace of the line, and they do not interact as a group. An assembly cell requires the following minimum conditions to be met (Burbidge 1989, Baudin 2002):

- Continuous flow. Products are assembled one at a time in a common sequence.
- A set of co-located assembly stations.
- A team of workers operating the stations.
- Self-contained facilities. The workers have all the tools and equipment needed to complete assembly and associated tasks and have full control over the facilities.
- Autonomous pacing. Workers control the pace of the work.

There is a considerable body of literature, especially the socio-technical literature, that clearly shows that assembly cells perform better than traditional, long, machine-paced assembly lines. In developed economies, workers will not willingly accept working on a traditional assembly line and overall productivity will be low. It might be thought that these problems can be overcome by performing manual assembly in developing and emerging economies. Even here, though, scientific studies have shown that groups perform better than individuals on a machine-paced assembly line, e.g., the early studies by Rice (1958) in India. The socio-technical systems approach is discussed in Chapter 14. Here we focus on the technological requirements for effective design of assembly and test cells.

There are three main ways to assign products for assembly (Baudin 2002). High-volume products will be assembled in dedicated cells, single-product cells. Medium-volume products are grouped based on similarity of parts and assembled in family product cells. Low-volume products are assembled in generic assembly cells.

There are four main ways to arrange assembly cells in a factory (Burbidge 1989): mono-group, parallel group, branched group and serial group. A mono-group assembles one product or family of products. If there are several mono-groups, then each assembles different products or product families. Parallel groups assemble the same products/product families. A branched group is one in which several groups assemble sub-assemblies, which are then joined together by another group in final assembly (converging branched group) or in which a sub-assembly group divides into different assembly groups (diverging branched group) as depicted in Figure 4.3. A serial group consists of a number of groups in a series, each group completing part of the assembly process. It is possible to combine groups; for example, a converging branched group may be followed by a diverging branched group.

There are three main ways to arrange assembly work within a cell (Burbidge 1989): mono working (individual build), team working (group build) and line

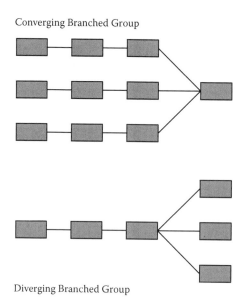

Converging Branched Group

Diverging Branched Group

FIGURE 4.3
Branched groups.

working. One person can easily assemble small products, such as electrical appliances. Thus there is one station for each person and assembly is completed at that station. Several people can be grouped together to form a cell, with several stations. This is called mono working. The workers in the cell may make the same or dissimilar products. The product does not move within the cell. This is a minimal group.

Large products generally require several workers working together at the same time to complete all or part of an assembly. This is team working or group build. Normally the product is assembled in one location (one station). Line working consists of several assembly stations connected together in series.

The procedure for designing assembly cells is:

1. P-Q analysis. The purpose of P-Q analysis is to sort products according to production volume. The resulting information is used to assign products to assembly cells: dedicated cells for high-volume products, product family cells for medium-volume products, and generic cells for low-volume products (Baudin 2002). A typical P-Q chart is shown in Figure 4.4. The quantities are given in number of products per month.

2. Factory layout. Determine the types and number of cells in the factory and arrange the layout in the factory.

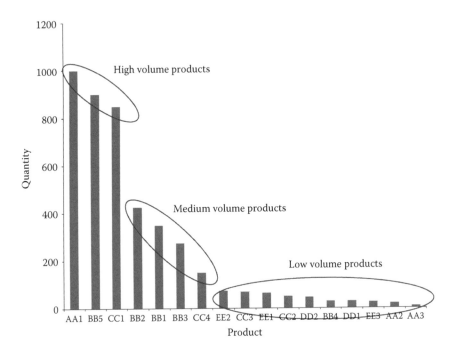

FIGURE 4.4
P-Q chart example.

3. Cell layout. Determine the best type of arrangement within each cell and design the overall cell layout.
4. Station layout. Design the layout for each workstation in a cell.

The following design guidelines apply to assembly and test cells (Burbidge 1989, Baudin 2002).

Cells

- Assignment of products to cells. High-volume products should be assembled in dedicated, single-product cells. Medium-volume products should be assembled in family product cells. Low-volume products should be assembled in generic cells.
- Group technology interface. If group technology is used to manufacture parts, then the assembly cells can be matched to the GT groups. That is, the families of products assembled by an assembly line or cell should match the families made in the GT cells. This reduces the lead time.
- Cell boundary definition. Assembly stations and equipment should be located close together and clearly separated from other cells.

- Task protection. Ensure that once materials enter an assembly cell they can be processed wholly within the cell without external disruption.
- U-shape layout. Use U-shape layout for line working, with the workers on the inside and parts feeding from the outside.
- Group build. Group build is appropriate for large products.
- Job enlargement. Include preparatory, support and maintenance tasks in the cell. That is, the workers should be accountable for all tasks needed to operate the cell.
- Autonomous pacing. Workers should be allowed to control the pace of work.
- Multifunctional workers. Workers should be trained to operate all stations in the cell and be able to undertake other tasks, such as maintenance.
- Water spiders. Include a water spider role in the team when using line working. The purpose of the water spider is to prepare parts and kits for the assemblers and to relieve them for short periods. The person should be the most expert assembler and cross-functionally trained in all the assembly positions.
- Small flexible machines. It is better to use small, flexible machines rather than high-output inflexible machines. This reduces the investment cost and minimises the economic batch size.
- Part presentation (kits). Deliver parts to the cell in assembly kits. The kits should be provided directly from suppliers. The kit containers are sent back directly to suppliers for re-filling and to initiate the order process.
- Part presentation (packaging). Minimise packaging. Where packing is required, remove packaging before delivery to the cells and ensure packaging can be recycled.
- Green space. Provide a seating area with a table inside the boundary of each cell, so workers can take rest periods. This area can be used for team development and problem-solving activities as well.
- Rework. Rework can be performed at the assembly cell or at an offline rework cell. The main advantage of performing it in the cell is that it places the rework responsibility on the assemblers who produced the defective product. This reinforces their accountability and provides them with opportunities to find ways to reduce defects. There are two major disadvantages of reworking in assembly cells. First, if an assembly line is used, it will be interrupted until the defective part is repaired. Second, disassembly tools have to be provided in the cell and workers trained in disassembly as well as assembly. The decision on where to locate rework must take into account these factors.

- Successive testing. The person or group assembling a product should not test it. Testing should be carried out in the cell where the product is assembled, but tested by an independent tester. It is better to test at the source than at a succeeding station or cell.
- Production planning and control. Use a JIT ordering system (pull system).

Stations

- Ergonomics. Design stations so workers are standing, not sitting. However, if workers are required to crouch, then it is better to provide flexible seats, e.g., Toyota's cantilevered raku-raku seats.
- Housekeeping. Use 5S to design each workspace in the cell.
- Tools. Attach tools to the workstation and not to the assembler.
- Visible management. Use assigned locations and orientations for storing tools. Use clear labelling. For line working, use tower lights, stop lights and other types of indicators to indicate cell status.
- Single-piece presentation. Parts may be supplied to the cells in racks. Single-piece presentation is a Japanese technique for optimising rack space. The requirements for single-piece presentation are (Baudin 2002, p. 187):
  - "One and only one unit is presented, with its smallest dimensions facing out.
  - Once it is picked, the next unit falls into place automatically.
  - Replenishment takes place from behind, without disrupting production."

Global Manufacturing

Germany was the world's leading exporting nation in the decade preceding 2009, when it was overtaken by China. Germany exports more than the United States and has done so for over a decade. What is astounding is that the majority of its exports are from small and medium-sized (Mittlestand) companies, not large multinationals. What is even more amazing is that many of these companies start exporting within a short period of their founding. There are many small and medium-sized companies in other countries that export as well. These companies are not well known and do not appear in the mainstream management literature. Hermann Simon (1996, 2009) has been studying them since the 1980s and calls them "hidden champions". Sixty-six percent of these companies are market leaders on the world market; 78% are leaders in Europe. They are truly global market champions.

To qualify as a hidden champion a company must be number 1, 2 or 3 in the world market, have revenue below US$4 billion, and a low level of public awareness. The revenue figure is very high; however, approximately one-quarter of the companies have revenues less than US$70 million. Sixty-nine percent of the companies are in industrial goods, 20% in consumer goods and 11% in services. Thirty-six percent are in engineering. Hence they are excellent models for small and medium manufacturing companies.

These companies are superb survivors. The oldest was founded in 1452, and one-third are over 100 years old. Fifteen percent are less than 25 years old, and yet they are world market leaders. Hidden champions achieve world dominance because they serve niche markets; many are one-product, one-market companies. Around two-thirds of the hidden champions are from Germany, Austria and Switzerland. The rest come from all over the world. Table 4.1 lists a sample of hidden champions, taken from Simon (2009, pp. 2–13).

Ponsse is a Finnish company that manufactures forestry machines and equipment. Ponsse is number 2 in the world after John Deere. Ponsse is not included in Simon's book, but it is also a hidden champion. A case study of the company appears in Chapter 13.

Simon's (2009) lessons learned from the hidden champions are summarised below as design guidelines for global manufacturing. There is one

TABLE 4.1

A Sample of Hidden Champions

Company	Country
Orica	Australia
Embraer	Brazil
Petzl	France
Arnold & Richter	Germany
Baader	Iceland
Essek Propac	India
Saes Getters	Italy
Ulvac	Japan
Gallagher	New Zealand
Tandenberg	Norway
Amorin	Portugal
International SOS	Singapore
Sappi	South Africa
Lantal	Switzerland
De La Rue	United Kingdom
Bobcat	United States

essential prerequisite: strong leadership with clear, long-term goals. Global markets are risky and the road is bumpy. Patience and tenacity are needed to succeed.

- Market focus. Focus on niche markets. Make one or a few products and excel at producing and selling them to a narrow market. Define the market by application and customer needs or by type of customer (target group). Define the market boundary in a way that suits the company's strengths. Do not follow standard definitions of markets.

- Depth. Vertically integrate along the supply chain. Develop core strengths and excel at them. This combined with a narrow market focus enables small companies to compete internationally.

- Decentralisation. Decentralise wherever possible. For small companies this means that staff should take on multiple roles and be given accountability for producing results. Larger companies can spin off smaller subsidiaries globally. All units should be small.

- Globalisation. Use globalisation to grow. Niche markets, by definition, are small. A home market may be too small to support a company, but a world market can. Enter world markets early, especially those where there are no competitors. Establish wholly owned subsidiaries in the world markets. These should be factories or technical service centres. If your long-term target is to export to the United States, Europe and Asia, then consider locating the HQ in Europe. This is the best global location for reaching all markets. Consider exporting to the emerging economies as these are growing rapidly.

- Innovation. Innovate to maintain technological leadership. Innovations should be driven by market and technology. Internally, top management must stimulate and drive innovation. Employ high-quality specialists in R&D and design. The quality of employees is more important than budgets. Use top customers and competitors to improve products and services.

- Closeness to the customer. Develop and maintain very close relationships with all customers, but especially the most demanding ones and those who are technology leaders. "Closeness to the customer is more important than marketing professionalism" (Simon 2009, p. 356).

- High-performance employees. Demand and create a culture of high performance with low tolerance for shirking. Keep the organisation lean; there should be more work than people. Jobs should be designed so employees can undertake multiple roles and flexible work assignments. Provide training and continuing education to employees.

Exercises and Problems

1. Briefly describe the four types of discrete part manufacturing.
2. Discuss the relevance of group technology for SMMCs.
3. List and briefly describe the main steps in production flow analysis.
4. List and briefly describe the steps in setting up assembly cells.
5. Where should test cells be located? Explain your answer.
6. List the design guidelines for global manufacturing.

References

Arn, E.A. 1975. *Group technology*. Berlin: Springer-Verlag.

Baudin, M. 2002. *Lean assembly*. New York: Productivity Press.

Burbidge, J.L. 1971. *Production planning*. London: Heinemann.

Burbidge, J.L. 1975. *The introduction of group technology*. London: Heinemann.

Burbidge, J.L. 1979. *Group technology in the engineering industry*. London: Mechanical Engineering Publications Ltd.

Burbidge, J.L. 1989. *Production flow analysis*. Oxford: Clarendon Press.

Gallagher, C.G., and Knight, W.A. 1973. *Group technology*. London: Butterworth.

Hayes, R.H., and Wheelwright, S.C. 1979. Link manufacturing process and product life cycles. *Harvard Business Review*, 57(1), 133–140.

Miltenburg, J. 1995. *Manufacturing strategy: How to formulate and implement a winning plan*. Portland, OR: Productivity Press.

Reinertsen, D.G. 1997. *Managing the design factory*. New York: Free Press.

Rice, A.K. 1958. *Productivity and social organization: The Ahmedabad experiment*. London: Tavistock Publications.

Simon, H. 1996. *Hidden champions*. Boston: Harvard Business School Press.

Simon, H. 2009. *Hidden champions of the 21st century*. Dordrecht: Springer.

Singh, N., and Rajamani, D. 1996. *Cellular manufacturing systems: Design, planning and control*. London: Chapman & Hall.

Slack, N., Chambers, S., and Johnston, R. 2004. *Operations management*. Harlow, UK: Pearson Education.

Smith, P.G., and Reinertsen, D.G. 1998. *Developing products in half the time*. New York: John Wiley.

Wilhelm, B. 1997. Platform and modular concepts at Volkswagen—Their effect on the assembly process. In K. Shimokawa, U. Jurgens, and T. Fujimoto (eds.), *Transforming automobile assembly: Experience in automation and work organization*, 145–156. New York: Springer-Verlag.

Section II

Context

Prologue to Section II

The business environment provides both opportunities for and constraints on small to medium manufacturing companies (SMMCs). For this reason environmental analysis is an important aspect of business planning. The business environment can be analysed in different ways. One common descriptive technique is referred to as PEST, which classifies the environmental characteristics under political, economic, social and technological. This original method of analysis has been extended to include legal and environmental aspects to give PESTLE. In Europe, the research community uses STEEP; they do not have a separate category for legal. More recently some business analysts have added ethics to PESTLE to give STEEPLE. The economist Ghemawat (2011) proposes a framework based on distance called CAGE: culture, administration, geography and economics.

Analysis of the business environment only has meaning in terms of progress. In Chapter 2 we showed that progress can be measured in terms of advancement toward the four ideals of plenty, good, truth and beauty. Hence we believe the business environment should be analysed in terms of the four ideals. None of the above techniques covers all four ideals, as can be seen in Table P.1.

We propose an extension to the European research community STEEP, called EASTEEP, by adding ethics and aesthetics. EASTEEP stands for ethics, and aesthetic, social, technological, environmental, economic and political contexts. EASTEEP covers the four ideals (Table P.1).

You may note that we arranged the letters so the acronym starts with EAST. Our aim is to highlight the growing influence of the East and emerging economies.

TABLE P.1

Comparison of Contextual Analysis Techniques

Business Environment Analysis Techniques				
PEST	**CAGE (Ghemawat)**	**STEEPLE**	**EASTEEP (Authors)**	**Ideals**
Political, economic, Social	Culture, administration, economic	Political, economic, social	Political, economic, social	Plenty
—	—	Legal, ethics	Ethics	Good
Technical	Geography	Technical, environmental	Technological, environmental	Truth
—	—	—	Aesthetic	Beauty

The chapters forming Section II briefly discuss the global manufacturing environment in terms of the EASTEEP framework and provide guidelines for regenerative development. Regenerative development refers to small communities supporting local ecologies. The approach stresses the interdependence of humans and nature and the need for humans to adopt development strategies that enhance the evolution of natural ecologies.

> Regenerative Development enhances the quality of ecosystems and human settlements by improving the ecological, cultural and economic health of a place. Humans must be co-developers with the evolving natural systems. Rather than merely reducing ecological degradation, Regenerative Development improves the health of habitats, the strength of social networks, and the depth of a community's historical roots. This perspective is the essence of thriveability: enhancing natural systems and supporting flourishing social networks. The approach recognizes the interdependence of humans and nature and the mutual benefits of conscious, responsible, co-developing interactions. (Edwards 2010, p. 76)

Regenerative development is a much stronger concept than sustainable development. The term *sustainable* implies a limit to development, whereas the term *regenerative* implies endless evolution. We are propounding the adoption of the regenerative development approach in this book.

References

Edwards, A.R. 2010. *Thriving beyond sustainability.* Gabriola Island, BC: New Society Publishers.

Ghemawat, P. 2011. *World 3.0.* Boston: Harvard Business Review Press.

5

Ethics

What Is Morality?

Morality is the distinction between right and wrong. The basis for the distinction can be found in religion, philosophy or culture. People can act morally without necessarily thinking about their moral behaviour. On the other hand, they can think about their moral behaviour; this is called ethics.

Other people judge us according to their moral principles whether we like it or not. The judgements may appear harsh. It takes a lifetime to build up a good reputation, but only one mistake to destroy it. This is because morality is not a goal that can be achieved. The qualities that define morality have to be affirmed and re-affirmed in our daily relationships with others. We act responsibly toward other people because they are important to us. Heyd (2007) refers to these targets of morality as *significant others*.

Moral people demonstrate their morality through their empathy, compassion and love of their fellow humans. Immoral people demonstrate their lack of morality through the exploitation of other people for their own selfish gains. Amoral people demonstrate their lack of morality through their indifference to the fate of other people; they simply could not care less about them. Moral people accept responsibility for their actions and act to maintain and improve their relationships with other people. Morality further requires that relationships are perceived for what they are, that there is no self-deception about them (Heyd 2007).

The basis for morality is how we treat significant others as individual human beings. This raises the question of whether we consider only the living, those people that we can interact with during our lifetime, or whether we should also consider those who lived in the past and those to come in the future. The experimentalists see morality as an "endless conversation of the living with our ancestors about how to design a good world for future generations" (quote attributed to C. West Churchman). The conversation is not a sterile debate. It leads to real action to change the world for the better. What does better mean? Or, in other words, how ought we live our lives? The experimentalists answer that one ought to pursue ideals. The endless progress toward the ideals brings contentment and makes life worth living.

"Singer made the theme of endless process a central one in his philosophy; his name for the restlessness he had in mind is 'contentment'…. What appear to be opposites, the restless and the contented, become the opposites of the same idea when we realize that 'contentment' comes from the Latin continere, 'to hold together.' The contented life is the complete life, made up of all those aspects of a life that make it meaningful" (Churchman 1961, pp. 199–200).

Experimentalist Ethics

Experimentalist ethics is a powerful, secular, social morality. It encompasses all aspects of life and all people: past, present and future. People are linked through cooperation to further progress toward omnicompetence as follows:

- Oneself: Each person should strive to gain more knowledge and understanding. Learning helps us to progress toward the ideal state of truth. What truth should be sought? Emery (1981) argues that the environment we live in is so uncertain that it is not possible to select means to use resources efficiently. Hence the truth is not knowledge and understanding about efficiency. Instead, he argues that we should aim to nurture resources for future generations, i.e., maintain and improve the resource pool. "The emerging ideal is that resources must be regarded as at all times part of the common pool of society's resources even though at any one time some individual or organization (public or private) had definite privileges of access to those resources…. One thing emerges clearly; the name of the ideal we are seeking is not that of 'Truth'" (Emery 1981, pp. 438–439). "The culture free ideal is best conceived of as the probable effectiveness of cultivating, not making" (Emery 1974, p. 62). We agree with Emery's suggestion to name it the *ideal state of nurturance*.

- Contemporaries: The living can help each other by providing resources and working together cooperatively to achieve more than one person can on his or her own. They help progress toward the ideal state of plenty. Plenty means plenty of cooperation, not plenty of resources. Cooperation between people is more important than having plenty of tools and resources (Emery 1974, 1981). We propose to name it the *ideal state of cooperation*.

- Past generations: The ideal state of omnicompetence has no meaning if all desire is lost. Humans are beings who have desires. In the ideal state humans must continue to desire, and this, in turn, "requires that more desirable goals and objectives replace old ones once they have been obtained…. The ideal that men pursue is not that they simply prefer what others prefer, but that they actually desire and affectively react in ways that are not necessarily the same as are those of others, but are not contradictory" (Emery 1974, p. 63).

Past generations provide us with emotional experiences that bind us as humanity. Art uses these experiences to create moods that inspire us to change the world and relax us when we tire of ideal pursuit. The aim of this ideal is to re-create the creator. Beauty is not the correct name for the ideal. We propose to name it the *ideal state of re-creation* (see Chapter 6 for more discussion).

Emery notes: "One implication of what we are postulating as an ideal is that men will increasingly reject the pursuit of purposes that are likely to be ugly, deforming, degrading or divisive" (1974, p. 64).

- Future generations: Future generations provide a basis for selecting goals and objectives by encouraging the living to consider the long-term consequences of their actions. They help progress toward the *ideal state of good*, where good is defined in terms of each individual being, not social groups or societies (Emery 1974, 1981). What is the good? It is peace on earth: no conflict within and between people. In practice, it means choosing purposes in such a way that their achievement helps other people achieve their purposes. This is the moral basis for all decisions. Maxwell (2003) expresses it as a golden rule. When you are making a moral decision or face a moral dilemma, ask yourself the question: "How would I like to be treated in this situation?" (p. 21). Chapter 9 will discuss an extension of the ideal state of good that encompasses non-human entities, such as animals, plants and natural landscapes.

  There is one further point to note. Exemplars of the ultimate good are models to be emulated and will be remembered; in this sense they become immortal. Such people demonstrate, by their actions, the fundamental goodness in human nature and encourage others to follow their example.

The ideals and their societal functions are summarised in Table 5.1.

How difficult is it to follow experimentalist morality, to be a progressive? "This answer begins by pointing out that though the sacrifices called for by the spirit of progress may be few in kind, yet they must be great in cost…. For to the class of ambitions the progressive is bound to fight in himself and others belong all those that can be fulfilled for one only by being defeated for some or all others. The classic image of a figure expressly created to portray the individual all of whose desires are subordinated to the will to overpower his neighbour is the Nietzschian 'Man of Might,' to whom the closest historic approximation is the Caesarian conqueror … but the same type of will on a smaller scale is plentiful enough in the world; it is to be found in any

TABLE 5.1

Ideals and Their Societal Function

Original Name (by experimentalists)	Revised Name	Societal Function
Ideal state of beauty	Ideal state of re-creation	*Aesthetic functions*: To intensify and purge emotions, to create dissatisfaction and satisfaction.
Ideal state of truth	Ideal state of nurturance	*Knowing functions*: To seek knowledge and understanding to nurture nature and humankind.
Ideal state of plenty	Ideal state of cooperation	*Politico-economic functions*: To improve cooperation among people to effectively develop and utilise earth's finite resources.
Ideal state of good	Ideal state of good	*Ethico-moral functions*: To remove conflict within and between people, and to live in harmony with nature.

man whose ambition is such that all men having the same ambition must necessarily stand in competition with him and with each other" (Singer 1948, pp. 32–33). In other words, it is very difficult. Our world is based on competitive struggle and not on cooperation (Mumford 1971, Kohn 1986).

Each of us has to make a moral choice about how to live. Some will choose a religious path, others a secular path. Whichever path you choose, experimentalist ethics can improve your moral decisions and help you to live a life of contentment.

Your moral behaviour determines who you are in your personal life, professional life and in business. There is no such thing as professional ethics and business ethics (Maxwell 2003, Heyd 2007). There is only ethics. However, the moral dangers and risks are often higher in professional life and business relationships. The remainder of the chapter discusses these and provides moral guidelines for professionals and SMMCs.

Ethical Aspects of Professional Life

There is a common expectation that professionals be held to a higher ethical standard than the rest of society. This viewpoint is based on a notion, however vaguely expressed, of a profession. Unfortunately, the term *profession* is used today to mean many different things. In fact, the word has been abused so much that most countries protect the use of it for certain occupations by law. For example, in Singapore the Professional Engineers Act specifically protects the use of the words *professional engineer*. Other protected professions include doctors, psychologists, lawyers, judges, clerics and accountants.

What is so special about those occupations called professions? The answer lies in a historical analysis of the development of the term (Barker 1992, pp. 84–86):

> This term came into use in late medieval times. The occupations that initially came to be called professions were medicine, the law, the clergy, and university teaching—four occupations for which study in the medieval university prepared people....
>
> Why did these occupations come to be called professions? Our term "profession" is from the Latin *"professionem,"* whose core meaning is that of a public declaration, but which in medieval times had come to mean the taking of religious vows.... The English noun "profession" in the thirteenth century means the declaration, promise, or vow made by one entering a religious order. Then, starting in the fourteenth century, it comes to mean any solemn declaration, promise or vow. It is only in the sixteenth century that it comes to mean an occupation in which learned knowledge is applied to the affairs of others, as especially in medicine, law, divinity, and university teaching.... In those times, the requiring of oaths and avowals, made publicly and in the name of God, was a powerful means for controlling human conduct and human attitudes....
>
> The term "profession" at first meant the battery of oaths, and then later by a natural extension came to mean the vocation into which the successful candidate entered after taking those oaths. This, I think, gives us the original core meaning of the term "profession" in its application to certain occupations requiring advanced study: the student who had completed his studies and was about to enter into the occupation of law-yer, physician, cleric, or university teacher was expected to *profess* his dedication to the distinctive ideals and practical realities associated with that occupation. The physician-to-be had to profess that he would use his medical knowledge to promote the bodily health of others, always for good ends (as the ancient Hippocratic oath affirmed). The would-be lawyer had to profess willingness to use his mastery of the law to promote justice. The person who was going to become a cleric had to profess that he would use his learning and sacramental powers to pro-mote the spiritual well-being of those with whose care he would be charged. Those who were going to become university teachers had to profess dedication to an educational ideal of service.

In short, the three main features of a medieval profession were:

- Mastery of an extensive body of knowledge achieved after many years of intellectual study and practical training
- Knowledge and practical skills that provided an important service in society
- Commitment to an ideal of service that imposed ethical standards more exacting than the cultural norms of the time

These essential features have carried over to the professions today and are encapsulated in the following definition of a profession.

A profession is a vocation that has the following features (Callahan 1988):

- Mastery of an extensive body of knowledge achieved through intellectual study at a university and practical experience at applying the knowledge.
- Knowledge and practical skills that provide an important service in society.
- An implicit or explicit social contract between the profession and society. The basis of this contract is a *professional-client relationship*. In the traditional social contract, a professional's obligations are to serve the client within the limits of the professional's special competence, to respect confidences obtained in the professional relationship, and not to abuse the powers to obtain undue advantage over or benefits from the client. A professional is expected to be committed to an ideal of service that is more ethical (more exacting) than current cultural norms. A professional is expected to abide by the ethical code of the profession irrespective of conditions imposed by employment contracts and even if no contract exists between the professional and the client. An example of the latter might be a doctor on holiday who attends to a person who has collapsed from a heart attack. The doctor is professionally bound by the medical ethical code to offer assistance even though no compensation may be given for providing medical services.

A society gives a profession freedom to run itself and professionals the right to make decisions affecting other people's lives. This is the right to practice the profession. In many countries this right is given and controlled by the law. A client's obligations are to accept the authority of the professional, to submit to the professional's advice and to pay for the services rendered.

A *professional association* is a group of members of a profession. The two main functions of a professional association are to ensure that the members serve the vital needs of a society in a proper manner and to protect and look after the welfare of its members. A society gives a professional association freedom to choose its own courses of action, goals and objectives in serving the vital needs of the society. With freedom comes accountability; that is the other side of the social contract. A professional association determines how its members ought to behave through a *professional code of conduct* and punishes those members who intentionally or negligently disobey. In some countries the code of conduct is determined by the government, e.g., Singapore. Non-compliance is a criminal offence and can result in imprisonment.

The traditional social contract has been criticised for two main reasons. First, it assumes the client in the professional-client relationship is one person or a small group of people. But the work of a professional can affect many people. How are their views to be taken into account? What happens if they disagree with the "client" who has commissioned the professional? Who will represent these people? Second, it gives professionals the right to make value judgements on behalf of their clients. The basis for this grant of rights is that professionals have special knowledge due to their training. Professionals have special technical knowledge (by definition), but many do not have special knowledge or training in making value judgements, e.g., engineering professionals.

Some researchers and practitioners, including the authors, argue for a new form of social contract in which the professional involves clients and other stakeholders in the decision-making process. The professional will be an expert in managing groups of people—the stakeholders—so consensus can be reached on value judgements. Section IV discusses how to organise manufacturing companies and design teams to facilitate stakeholder involvement.

In practice, following a professional ethical code of conduct can be very difficult. There are many opportunities to cheat, and there may be considerable pressure from peers and superiors to bend the rules. The downward slide to immorality does not occur in one big decision, but by many small steps; each step moving further and further from the moral norms of the profession. One way this can happen is through a type of immoral behaviour called wilful blindness (Heffernan 2012).

Wilful Blindness

Wilful blindness is a legal term. It refers to situations in which defendants plead innocence on the basis that they did not know what was going on, but the court holds that they could and should have known. Ignorance is not an excuse in situations where people could and should have ensured that they were not ignorant.

Wilful blindness is probably the most deadly immoral behaviour that professionals have to protect themselves against. Heffernan's book will alarm all professionals, especially as her arguments are backed up by well-authenticated psychological research studies. In essence, her argument is as follows. Our worldview defines our personality. When our worldview is threatened, then our whole personality, who we are, is also threatened. So we do whatever we can to protect ourselves. We ignore contrary views and discordant facts, and we find ingenious ways to rationalise our position. Some of the more common ways blindness can occur are (Heffernan 2012):

- Affinity. We select partners and friends who are similar to us. Their views reinforce ours and blind us to diverging points of view.

- Love. It is very difficult to see faults in people whom we love. Love is truly blind.
- Ideology. Strongly held convictions prevent contrary views from being considered and ensure discordant facts are ignored. Worse still, the more others point out the failings in our convictions, the more strongly we defend them.
- Limits of our minds. Psychologists have shown there are severe limitations to the number of tasks we can handle. We are very poor at multitasking. We become more and more limited in our focus as our minds become overloaded, whether due to stress, tiredness or other factors. More importantly, our morality declines with overload. For example, we are more likely to make less moral decisions when we are tired. There is a good reason for the 40-hour work week. Longer hours lead to tiredness and a reduction in performance and moral standards.
- Head in the sand attitude. We ignore what we do not want to hear. We see only what we want to see.
- Blind obedience: Following orders we know are morally wrong. The instinct to obey is very strong and difficult to overcome, especially if disobeying is likely to result in loss of one's job.
- Conformity. When we belong to a group there is tremendous pressure to conform to group norms. Failure to comply can lead to isolation and ostracism. So we go along with the group even though the members may behave immorally.
- Bystander syndrome. Studies show that our behaviour is influenced by what others do. In a crowd, we expect someone else to act in a situation requiring a moral decision. When no one does we follow suit—we do nothing when we could have. The syndrome is not restricted to crowds; it applies to companies as well.
- Out of sight, out of mind. The farther away we are from people geographically, the more unlikely we are to empathise with them. Skype and other forms of distance communication cannot make up for face-to-face discussions. As distances increase our empathy decreases and ethical standards drop.
- Economic man. This blindness occurs when we put a price on human life and happiness. Everything is reduced to a price: people, the environment, natural beauty, animals, etc. In Heffernan's words: "What money does … is allow us to disengage from the moral and social effects of our decisions. As long as we can frame everything as an economic argument, we don't have to confront the social or moral consequences of our decisions. That economics has become such a dominant, if not prevalent, mindset for evaluating social and political choices has been one of the defining characteristics of our age" (2012, pp. 261–262).

Speaking out and challenging wilful blindness is dangerous. Whistle-blowers are disliked and ostracised, despite laws to protect them. Many are unable to gain employment in their profession or in the same country. The collective blindness of our societies is the greatest human tragedy of all time. Though we encourage professionals to speak out, we must warn them that the sacrifices can be huge. We advise professionals to resign if they work for companies that act immorally and they are unable to take the big step to blow the whistle.

Ethical Guidelines for the Professional

Professional ethical rules of behaviour are written in codes of conduct. Codes of conduct are necessary but not sufficient for moral behaviour. They fail in practice, for several reasons (Johnson and Riley 2008, p. xvii):

> (1) Ethical quandaries are complex and their resolution often nuanced and resistant to simple rules; (2) Ethical quandaries are fluid and demand considerable flexibility on the part of the professional; (3) Professionals often encounter competing obligations to individuals, organizations, and society at large, and there are often good reasons for different courses of action; (4) Being ethical is a continuous process, not merely a right answer; and (5) Ethical decisions are made by fallible human beings, many of whom are motivated by self-interest, defensiveness, and a remarkable capacity to justify unethical courses of action.

A professional is expected to act on *ethical principles* and *moral virtues*. Ethical principles are rules on what to do. Moral virtues determine who you are. Both are required. Ethical guidelines based on experimentalist morality and moral virtues based on Johnson and Riley (2008) and Maxwell (2003) are given below.

General Ethical Principles (What You Should Do)

- Professionals with religious beliefs will uphold the best moral principles of their religions, be tolerant of other religions and secular beliefs, and should consider adopting experimentalist ethics to improve their moral behaviour.
- Professionals who use reason to develop a moral philosophy of life should consider adopting experimentalist ethics.
- All professionals will uphold the minimum standards of their profession as expressed in its code of conduct, irrespective of whether they are members of a professional society or not.

Moral Virtues (Who You Should Be)

- Maintain integrity. Integrity means being congruent, consistent and transparent in your work. Baltasar Gracián, a 17th-century Jesuit priest, expressed this point eloquently (Kaye 1992, p. 35). "The world is in chaos. Honorable dealing is deteriorating, good friends are few, truth is held in disrepute, good service is underpaid, poor service is overpaid. Whole nations are committed to evil dealings: With one you fear insecurity, with another inconsistency, with a third, betrayal. This being what it is, let the bad faith of others serve not as an example, but as warning. The real danger of the situation lies in the unhinging of your own integrity: accepting less than your best, being overly tolerant of stupidity, forgiving incompetence, fraternizing with the nonspiritual. The man of principle never forgets who he is, because he clearly sees what the others are."

- Treat others as you would want to be treated yourself. Specifically:

  - Do no harm to others. This involves "remaining alert to negative outcomes, resisting coercion, maintaining role boundaries, and remaining alert to the lure of sex, money, prestige and power. Ethical professionals refuse to exploit others and resist efforts by others to misuse their work" (Johnson and Ridley 2008, p. 37).

  - Be respectful. All people should be valued as human beings. They should be accorded dignity, which means differences in values should be respected, their privacy should be protected, and they should not be undermined.

  - Be compassionate. The essence of this guideline is to care for people who are vulnerable or less fortunate. Professionals should protect those who are vulnerable and volunteer their services for free to help the less fortunate. It requires an active stance, not a passive one. The professional needs to search for the less fortunate and vulnerable and take appropriate action.

  - Be just. Treat all stakeholders equitably. The rights of vulnerable stakeholders should be upheld. You, the professional, may be the only one who can speak up for them. It is your duty to do so.

  - Be faithful. Tell the truth and only make promises you can keep.

- Be beneficent. "Ethical professionals demonstrate flexibility and collaboration in their work—particularly when misunderstanding and conflict arise. They clarify expectations, communicate clearly, and achieve an ongoing sense of informed consent in their work. Ethical professionals make sure they are competent. They operate only within the bounds of their established competence and make referrals when new demands or requests exceed their capacity to perform competently" (Johnson and Ridley 2008, p. 79).

- Be prudent. Think carefully before acting, be cautious. Make sure that what you do is right and seen to be right.

- Promote autonomy. Professionals should respect and promote the autonomy of their stakeholders. As far as is feasible, stakeholders should be involved in decision making, and they should be fully informed of the consequences of their decisions. Professional designers should include stakeholders in the design process (see Section IV on how to do this).

- Do your best at all times. "An abiding commitment to deliver one's best is a hallmark of the ethical life" (Johnson and Ridley 2008, p. 175). In short, excel at all you do. Do not perform just to meet minimum standards or for legal compliance. Place the interests of your stakeholders first no matter how inconvenient this may be for you.

- Be aware of moral hazard. Moral hazard is where a person takes risks that he or she would normally avoid because there are no penalties or negative consequences for failure, but there are incentives if the risk works in that person's favour. "Moral hazard played a significant role in the recent economic crisis. In the securitization food chain, a mortgage broker who knowingly brought a liar loan to a bank got compensated for his efforts but bore no responsibility for what would happen as the mortgage moved down the line. Likewise, the trader who placed enormous bets on a CDO [collateralised debt obligation] would be rewarded handsomely if he succeeded but was rarely punished" (Roubini and Mihm 2011, p. 68). Resist moral hazard in your own behaviour, encourage others to resist it also, and take steps to change work situations that encourage moral hazard.

- Be aware of wilful blindness. Be vigilant and honest in assessing your own blindness and that of your work associates. Work actively to minimise it.

Ethics and Values in Business

The discussion, so far, has focused on a personal viewpoint of ethics. There is also a company viewpoint. A SMMC needs to set and enforce moral norms and values that determine its culture. Its culture, in turn, creates the identity of the SMMC that uniquely defines it. The identity and culture of a SMMC are created by the SMMC's stakeholder relationships. These can be defined in a statement of purpose. You may also note that a statement of purpose is an essential part of an idealised design of a SMMC, which is discussed in Chapter 15.

The key features of a statement of purpose are:

- It identifies the business the SMMC wants to be in. This may differ from the current business. The business should be defined as broadly as possible, but be precise.
- It identifies the major stakeholders and how the SMMC will relate to them.
- All major stakeholders, or their representatives, should be involved in the formulation of the statement of purpose as much as possible, as they have a critical influence on the effectiveness of its implementation. Employees can participate through the circular organisation structure (Chapter 15). The statement should be agreed upon by consensus as far as possible.
- It will be reasonably brief. An explanatory document can accompany it to aid interpretation.
- It is used as the basis for legitimising the SMMC's goals and objectives.

An example statement of purpose is shown below. It was developed for a primary school in 1989. Most of the key stakeholders were actively involved in developing it: board of governors, senior management staff, teachers and other staff, pupils and parents. It served as the guide for changing the learning environment for the children and working conditions for staff.

> The purpose of X Primary School is to provide educational services to children. More specifically, X Primary School will:
>
> - educate CHILDREN so that each child may develop independence and grow intellectually, physically, creatively, socially, emotionally, and morally to his/her maximum potential within a caring environment;
> - provide rewarding employment and professional growth opportunities for TEACHERS AND SUPPORTING STAFF which will enable them to develop towards their full potential;
> - encourage PARENTS/GUARDIANS of children attending the School to participate in the education of their children;
> - co-operate with other EDUCATORS, including parents, to ensure that children have a positive experience as they enter and leave the School, and to learn from these educators;
> - co-operate with the COLLEGE OF EDUCATION by:
>   - (a) being perceived as a client that appreciates its services and to learn from it, and
>   - (b) providing practical experience for student training;
> - be perceived by SUPPLIERS of goods and services as a client that appreciates them and to learn from them;

- be perceived by GOVERNMENT to be implementing the School's Charter and Government policy on education;
- co-operate with OTHER SCHOOLS for mutual benefit;
- (a) aim to be held in high regard by the LOCAL COMMUNITY, (b) make resources available to local groups for mutual benefit, and (c) act as a focal point for the local community in an emergency.

The statement of purpose can be transmitted and its implementation controlled by the following techniques:

- Making policies to support it
- Designing incentive schemes to support it
- Leading by example
- Educating and training staff in accordance with it

Furthermore, SMMCs can help progress toward the ideal state of good by actively striving to reduce or remove conflict within and between their stakeholders. Some conflict between employees can be removed through good job design (Chapter 14). SMMCs can also provide access to counselling services for employees suffering from stress and other psychological disorders. Potential conflict between stakeholders and a SMMC can be reduced through interactive planning (Chapter 15).

The remaining chapters in Section II provide guidelines for regenerative development. This opening chapter of Section II is important because it underwrites the other chapters. The following chapters are only meaningful if you, the reader, adopt an ethical approach that embraces all aspects of life. We provide the guidelines. You must provide the motivation to apply them.

Exercises and Problems

1. Compare your personal ethics with experimentalist ethics.
2. Two women in Thailand discovered that their bank had inadvertently transferred Bt2 million to their account by mistake (Kitlertsirivatana 2012). They returned the money. The bank denounced the women for undermining its credibility. Their neighbours said they were stupid for returning the money.
   a. Should the women have returned the money?
   b. Should the bank have denounced them?
   c. Were they stupid to return the money?
   Give reasons for your answers.

3. A fellow colleague, A, is fired from the company you both work for due to poor performance. Her boss, B, writes a letter of praise to assist A in getting another job. You query B about this and he replies it is just a small white lie. Discuss B's behaviour and his justification from a moral viewpoint.

4. In Australia, companies must provide health care insurance to their employees or they will be fined. The directors of one company calculated that it was cheaper to pay the fine each month than it was to buy health care insurance so they terminated insurance for all their employees. Is the directors' action moral or immoral? Justify your answer.

5. A professional photographer, covering a civil war, takes a photo of a young child running and screaming in pain from severe burns resulting from an incendiary (napalm) bomb that was dropped on her village. Should the photographer have assisted the young child? Explain your answer.

   (After you have answered the question go to http://www.dailymail.co.uk/news/article-2153091/Napalm-girl-photo-Vietnam-War-turns-40.html and find out what the photographer actually did and what eventually happened.)

6. Other than authorising professional associations, how could a community have its needs met for activities that require a long period of education, training and experience to perfect and provide a vital service to it?

7. Write a statement of purpose for the department or college where you are currently studying or for a university club that you belong to. Keep a log of the difficulties you faced when writing it.

References

Barker, S.F. 1992. What is a profession? *Professional Ethics*, 1(1–2), 73–99.
Callahan, J.C. (ed.). 1988. *Ethical issues in professional life*. New York: Oxford University Press.
Churchman, C.W. 1961. *Prediction and optimal decision*. Englewood Cliffs, NJ: Prentice-Hall.
Emery, F.E. 1974. *Futures we're in*. Canberra: Centre for Continuing Education, Australian National University.
Emery, F.E. 1981. The emergence of ideal-seeking systems. In F.E. Emery (ed.), *Systems thinking*, 431–458. Vol. 2. Harmondsworth, UK: Penguin.
Heffernan, M. 2012. *Wilful blindness*. New York: Simon and Schuster.
Heyd, T. 2007. *Encountering nature: Toward an environmental culture*. Aldershot, UK: Ashgate.

Johnson, W.B., and Riley, C.R. 2008. *The elements of ethics*. New York: Palgrave McMillan.

Kaye, J.L. 1992. *Practical wisdom for perilous times: Selected maxims of Baltasar Gracián*. London: Aquarian Press.

Kitlertsirivatana, E. 2012. Bangkok needs its own dark knight. Letter to the Editor. *The Nation*, August 11.

Kohn, A. 1986. *No contest: The case against competition*. Boston: Houghton Mifflin.

Maxwell, J.C. 2003. *There's no such thing as "business" ethics*. New York: Center Street.

Mumford, L. 1971. *The pentagon of power; the myth of the machine*. London: Secker and Warburg.

Roubini, N., and Mihm, S. 2011. *Crisis economics: A crash course in the future of finance*. London: Penguin.

Singer, E.A., Jr. 1948. *In search of a way of life*. New York: Columbia University Press.

6

Aesthetic Context

Introduction

The ideal state of omnicompetence is not a static state where all desires are satisfied. It is a state of alternating satisfaction and dissatisfaction. Herein lies a problem. It is so easy to fall by the wayside by being satisfied where you are and not progressing, or by being so dissatisfied with the perceived lack of progress you simply give up ideal pursuit. The purpose of the ideal state of re-creation is to make people dissatisfied when they are satisfied—inspirational purpose (intensifying emotions), and satisfied when they are dissatisfied—cathartic purpose (purging emotions). *The purpose is to re-create the creator.* "The artist creates no new ideals; he creates the creator of ideals" (Singer 1923, p. 40). Ackoff and Emery (1972, pp. 244–245) succinctly describe this ideal:

1. An ideal-seeking community must continually renew its members so that they are capable and desirous of striving for something better. But continuous striving for the unattainable exhausts men both mentally and physically. When mentally or physically fatigued, man can neither create new instruments nor conceive new means and ends. It is necessary, therefore, to provide him with a change of pace that he may be renewed, that the creator may be recreated. This is the *cathartic* function embodied in recreational institutions—sports, cinema, television, and so on. In art the comic is the cathartic form.
2. It is also necessary to inspire men to create and fashion new conceptions of the possible, and to implement them. Through beauty, art creates what may be called the *creative mood*, a disposition to create, and the *heroic mood*, a disposition to sacrifice present values for those of the future…. In art the tragic is the inspirational form.

   Art leads men to find new meanings and commitments in life, and the man that leads other men moves them with visions of the possible and desire for the unattainable. Thus leadership can be considered to be an art form.

Aesthetics is the philosophy of sensuous experiences (Fenner 2003). It includes experiences induced by art, nature and the environments we live in (home, gardens, cities, etc.). Artistic aesthetic experiences can be produced by static artefacts, e.g., a sculpture or painting, or by motion, e.g., a play, an opera, a dance or a musical performance. Beautiful artworks produce positive aesthetic experiences. Therefore philosophers have devoted time and effort trying to understand the meaning of beauty. They use the term *beauty* in two completely different ways (Lorand 2000):

1. It is used as a general term to refer to all aesthetic experiences.
2. It is used to refer to a measurement or a range of measurements of aesthetic value.

The experimentalists use beauty in the first sense. They also argue, as do many other philosophers, that ugly artworks can create positive aesthetic experiences. So now beauty includes ugliness. Lorand (2000) argues that beauty should be restricted to the second sense. We agree. The experimentalists' original name for this ideal (*ideal of beauty*) is not correct. We suggest it should be called the *ideal state of re-creation*. This directs attention to its purpose: to re-create the creator.

The Nature of Aesthetic Experience

Aesthetics is poorly understood. There is considerable disagreement about the meaning of beauty and what constitutes art and aesthetic experience. In this section we have selected characteristics of aesthetics that are important for product and manufacturing system design.

Figure 6.1 shows how *fine art* produces an aesthetic experience (Berleant 1997). First there is an artwork: an object (e.g., a painting) or the instructions for motion (e.g., the script for a play or a musical composition). The artwork is the object that induces the aesthetic experience. It is created by an artist. The artwork is intended to generate a *unifying perception*. If the artist is successful, an interested observer will be able to re-create the artist's unifying perception. In this sense, the artist is visible to the observer.

There is an interested observer (appreciator), who intends to appreciate the artwork. A successful artwork will engage the observer's attention long enough to produce sensations: these may be pleasurable or not. The sensations evoke emotions, which in turn generate a creative or heroic mood, or catharsis.

The one aspect not yet discussed is the performance of the artwork. "It is important to realize that, in a significant sense, all the arts are performing arts. A performer is not merely an agent who transmits the inspiration (or, if one

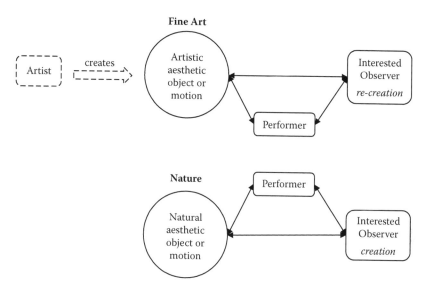

FIGURE 6.1
The nature of aesthetic experience.

is more prosaic, realizes the notation) of a composer or choreographer to an audience. In arts such as music, dance, and theater, the performer activates the art object, bringing it into perceptual play. Yet in the other arts, as well, the appreciator performs the same function in the dynamic appreciative process, realizing the aesthetic potential of the painting, film, or the poem through active perception" (Berleant 1997, p. 128).

The process is the same for natural aesthetic experiences, except there is no artist and the performer and interested observer are the same person.

There are several important points to note:

1. The interested observer is a free agent. He or she allows the evoked emotions to happen and must endure them long enough to obtain the aesthetic experience.

2. Aesthetic experience involves all our senses.

3. The interested observer immerses himself or herself in the experience. 'One is centered, perceiving things with enhanced acuteness and concentration. This is sometimes described as a magical moment in which the world becomes intensely vivid.... A powerful feeling of connectedness displaces the protective distance we usually impose between ourselves and the places we encounter, a distance not only physical but psychological" (Berleant 1997, p. 171).

4. All aesthetic experience is an act of creation (Berleant 1997, Arieti 1976). It is not only the artist who is creative, but the interested observer as well. The interested observer re-creates the unifying

perception of the artist for artworks and creates it for natural aesthetic experiences.

5. Each aesthetic experience is unique. Therefore aesthetic objects are not interchangeable. That is, they cannot be interchanged to give the same pleasure. This is one key distinction between aesthetic and functional objects. Each functional object is interchangeable in the sense that each is equally capable of performing the required function.

6. Aesthetic objects include social relationships.

7. All aesthetic experiences are social. First, the appreciative skill of the interested observer depends on his or her cultural background. Second, society determines what aesthetic objects are important by adding them to, removing them from, and preserving and maintaining our natural and cultural heritage. Society controls the aesthetic objects we can encounter and how we encounter them.

8. All environments have the potential to produce aesthetic experiences. In this chapter, we shall refer to the *aesthetic environment* to emphasise our focus on an environment's aesthetic potential. Whether an environment achieves its potential depends on its inherent aesthetic features and the ability and willingness of the interested observer to perceive its aesthetic quality.

Different fields in the philosophy of art have developed because philosophers have focused on different aspects of the process. Some philosophers have studied the characteristics of the artwork only. Others have studied the effect on the interested person. Still others have studied what the artist needs to do to create an artwork. There is so much disagreement about what constitutes art that some modern philosophers have defined fine art as what the art world (people and institutions knowledgeable in art) says it is (Levinson 2003). A more tautological definition is difficult to imagine! Notwithstanding this comment, the *art world view* is useful in that it highlights the role of societies and communities in developing and protecting the aesthetic environment.

Proponents of the art world view argue that the history of the object and artist must be known in order to have an aesthetic experience. This is not true for two main reasons. First, this would mean there is no way to evaluate new artists and novel art, as they have no history. Therefore there can be no aesthetic experience for new art. Second, it restricts aesthetic experience to a few knowledgeable experts. This ignores the fact that people have natural aesthetic experiences, where there is no prior knowledge. History and prior knowledge of the process that produces an aesthetic object are not necessary for an aesthetic experience (Heyd 2007). However, such knowledge may intensify the experience.

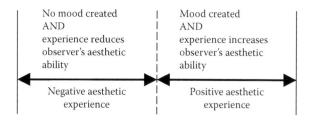

FIGURE 6.2
Range of aesthetic experiences.

There can be bad art as well as good art and negative aesthetic environments as well as positive ones. When we see an artist who is creating an artwork, the best we can say is he or she is *trying* to produce art. The artist might succeed in producing a positive aesthetic experience, a negative one or none at all. Berleant (1997) provides a basis for valuing aesthetic experiences, depicted in Figure 6.2. A positive aesthetic experience occurs if a creative or heroic mood or catharsis is achieved. The achievement improves the observer's ability to create further aesthetic experiences. A negative aesthetic experience occurs when no mood or catharsis is achieved and the resulting experience reduces the observer's ability to create aesthetic experiences. A neutral experience is one in which no mood or catharsis occurs and there is no impact on the observer's ability to create aesthetic experiences—it is represented by the dashed line in Figure 6.2.

Note that a positive aesthetic experience can be achieved through negative emotions. An ugly artwork may produce revulsion. A storm at sea may produce fear. But if the revulsion and fear lead to a creative or heroic mood, then they have generated a positive aesthetic experience.

Product and manufacturing system designers need to understand the range of experiences that can be created so they can avoid producing negative aesthetic environments and more successfully achieve positive ones. Some experiences are listed in Table 6.1.

Aesthetic Function

Some of the confusion about artworks arises because philosophers have sought beauty or aesthetic function in one artwork. This is not possible. The experimentalists very clearly view the aesthetic function as a function of the set of all artworks, as does Berleant (1997)—in this sense, they are in agreement with the art world view. The set of all aesthetic objects that we can perceive at a given time constitutes our aesthetic environment. It includes artworks, natural landscapes and features, everyday objects, homes, villages, towns, cities and urban and rural landscapes.

TABLE 6.1

Some Negative and Positive Aesthetic Experiences

Name of Experience	Aesthetic Effect (positive or negative)	Explanation
Ugly objects	Positive	Produces a feeling of revulsion but can still provide a positive aesthetic experience. Some ugly objects are necessary to provide contrast with beautiful ones.
Offensive	Negative	"Aesthetic offense diminishes us humanly by manipulating us perceptually in the interest of other ends, by exploiting our susceptibilities, by engineering measured anxieties to impair judgment, or by creating sheer discomfort" (Berleant 1997, p. 67).
Triteness, banality	Neutral	Fails to produce an aesthetic response. Banal art/environments also disappoint us because they do not utilise the full capabilities of the artists' craft. Nature cannot be trite or banal.
Dullness, boring	Neutral	Fails to produce an aesthetic response. In art, a dull or boring object/environment is a failure of the artist. In nature, it is a failure of the interested observer. Nature is not inherently dull or boring.
Inappropriateness	Neutral	The aesthetic object is not suitable for its context.
Trivialisation	Negative	The aesthetic object trivialises real and important values and issues by reducing the aesthetic experience to gratuitous sensations.
Deception	Negative	The artist creates the object/environment so that it appears to be something else. Art forgeries are included in this category. Other deceptive examples include theme parks, theme hotels, and housing developments and the use of plastic materials manufactured to look like wood or other natural materials. Deception is particularly bad because it "undermines people's sense of reality. What is environmental reality when the only thing that is real in people's experience is the false?" (Berleant 1997, p. 73).
		Note that this form of deception is different from the deception practiced in theatres for plays. Here the audience knows it is attending a play and that the props are not real. The negative artistic deception, on the other hand, refers to situations where the observer is deceived without knowing it.
Aesthetic harm	Negative	Aesthetic harm "coarsens perceptual consciousness, constricts the development of sensory awareness and the pulsating vitality of the body, and promotes sensory depravity" (Berleant 1997, p. 75). It includes all forms of pollution (Chapter 9).

TABLE 6.1 (*Continued*)

Some Negative and Positive Aesthetic Experiences

Name of Experience	Aesthetic Effect (positive or negative)	Explanation
The sublime	Positive	An extremely intense aesthetic experience of "power and magnitude so overwhelming that it cannot be circumscribed" (Berleant 1997, p. 77), e.g., gazing at the stars or being in a boat in a storm at sea. The sublime creates a feeling of awe and powerlessness.
Negative sublime	Negative	Large-scale systems and processes created by humans that are aesthetically degrading, e.g., weapons of mass destruction, mass marketing, mass culture, etc.
The sacred	Positive	An extremely intense aesthetic experience in which the sense of self vanishes.

Source: Berleant, A., *Living in the Landscape: Toward an Aesthetics of Environment*, Lawrence: University Press of Kansas, 1997.

The overall function of an aesthetic environment is to create a creative or heroic mood, or catharsis. Each object in the environment can contribute to this in various ways (Berleant 1997). At the simplest level is *mechanical function*: an ideal of perfect utility derived from the perfect machine. The aesthetic aspects are those of a machine: accuracy, exactness, precise operation, dependability, reliability, simplicity, etc. "It is an ideal embodied in the synthesis of production and perception, like the functional operation of the sailing vessel and the airplane, the smooth workings of the watch and engine, good form in the racehorse, and perfect timing in the trapeze act. The aesthetic of mechanical function finds its widest fulfillment in design, in architecture, and in city planning" (Berleant 1997, p. 88).

Organic function is the ideal embodied in the notion of system: of parts working together to provide higher-order (emergent) properties. The human body is the model for organic function in artwork, with its highest and most intense expression in dance. It includes mechanical function.

Practical function is pleasure derived from the use to which objects can be put. It is the aesthetic experience of everyday objects and environments. It includes both mechanical and organic functions.

The first three functions are incomplete. They can be considered functional steps toward the ideal of re-creation. *Humanist function* as described by Berleant (1997) is the ideal. "Here function becomes active participation, combining the mechanical, organic, and practical aspects of the art object and the aesthetic perceiver into a living movement of intrinsic, primary experience. From this, a synthesis of aesthetic perception, social relevance, and human fulfillment develops into a cultural environment in which each of these not only encompasses the others, but also becomes inseparable from them" (Berleant 1997, p. 94).

Environmental Aesthetics

Philosophers virtually ignored nature as a source of aesthetic experience until a seminal paper by Ronald Hepburn in 1966 (Levinson 2003). Since then a new field has developed: environmental aesthetics (Berleant 1997, Heyd 2007, Levinson 2003). Environmental aesthetics covers the aesthetic appreciation of all human environments, natural and otherwise. Heyd (2007) focuses on natural environments and approaches the field from an anthropological point of view, by studying how people actually relate to nature. One of the most difficult aspects of any aesthetic experience is remaining focused on the aesthetic object/motion. Heyd calls this aesthetic endurance. One practical way to improve endurance in natural environments is by learning stories about them. The stories guide and mediate the appreciation of nature.

Stories can be verbal artistic (e.g., poetry), verbal non-artistic (e.g., Australian aboriginal dreamtime stories) and non-verbal (rock art, tombs, engravings, paintings, sculptures, etc.). The purpose of the stories is to help the interested person view the aesthetic object in different ways. You may recall in Chapter 3 that alternative views are a source of inspiration for design creativity (Aspelund 2010). Using stories to appreciate nature can also be a source of inspiration.

Another practical way to appreciate nature is by wandering (Heyd 2007). Wandering is human-powered activity—walking, hiking, climbing, cycling, canoeing, etc.—for *its own sake*. The aesthetic pleasure comes from:

- *The activity itself (bodily motion in space).* Heyd sees wandering as similar to dance. Intrinsic satisfaction is gained from developing skills in body motion (the observer is the performer). In wandering these skills come from adapting one's body motion to different aspects of the terrain.
- *The ordering of objects and spaces along the journey.* Objects and spaces are encountered in an order as the journey progresses. Reflection on this order imbues new meanings on natural landscapes.
- *The unmediated appreciation of the surroundings using all senses.* All faculties are involved because there is direct interaction between the interested observer and nature, and this intensifies the sensations and, hence, the aesthetic experience.

Nature can be appreciated through indigenous rock art and reclamation art (Heyd 2007). Reclamation art is the reclamation of degraded parts of the environment as artworks. Rock art helps us see nature through the eyes of indigenous people. Research in Australia has shown that the aboriginal people have a rich and deep understanding of the stars and other astral phenomena, some of which is expressed in rock art (Norris and Hamacher 2010). One example is the emu in the sky (Figure 6.3), where aboriginal people

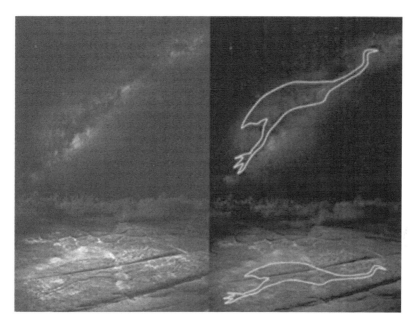

FIGURE 6.3
Emu in the sky. (Photograph courtesy of Barnaby Norris. Reprinted with permission.)

see the dark clouds in the Milky Way as an emu (an Australian bird). Figure 6.3 shows the emu in the sky at the top and an aboriginal rock carving of it at the bottom. Alternatively, South American indigenous people see a llama or ostrich in the sky. In the West we have inherited our knowledge of the skies from the ancient Greeks and Romans. It is easy to ignore the fact that other people will identify different patterns. By viewing the night sky through their eyes we can enrich our appreciation of it.

Aesthetics of Product Design

Product design produces objects that have instrumental and aesthetic value. A good design will be fit for purpose and aesthetically pleasing, i.e., have good practical function. These two aspects are complexly interrelated, which is why good design is so difficult. Worse, there is no universal agreement on what a good design is, but we can list some of its characteristics to achieve humanistic function (Bürdek 2005, pp. 15–16):

- Good design … must express the individuality of the product in question through appropriate fashioning.

- It must make the function of the product, its application, plainly visible so that it can be understood clearly by the user.
- Good design must allow the latest state of technical development to become transparent.
- Design must not be restricted just to the product itself; it must also take into consideration issues of ecology, energy conservation, recyclability, durability, and ergonomics.
- Good design must take the relationship between humans and objects as the point of departure for the shapes it uses.

Designers have developed different design guidelines for combining function with aesthetic features (form). Some generate form from function, others function from form, and still others take an iterative approach where neither form nor function takes precedence. To be successful, a designer must be familiar with the different guidelines and styles. Designers can use Alexander's (2002) design guidelines (Chapter 3) to help them design beautiful products and manufacturing systems.

Product designers face four major problems that artists do not. First, the design and its production are separated. An artist manually crafts the artwork. Product designers prepare a design, and then someone else is responsible for the manufacture. In fact, there may be little human input into the manufacture at all; most of it may be performed by machines. The result is a separation between designer and user. The designer has lost his or her identification with the product. Second, small to medium manufacturing companies (SMMCs) often have a product range, rather than a single product. In this situation, a SMMC will want the products to have the same look and feel so they can be identified with the company. Both problems relate to identification. Product designers have developed identification guidelines based on communication theory and language (e.g., semiotics). Bürdek (2005) is a good reference source.

Third, the product has to be designed in such a way that a user knows what it is for and how to use it (Bürdek 2005). Kinesthetic, visual and audio cues are very important. The provision of these cues may detract from, be neutral to or enhance a product's beauty. In additive design the cues are arranged so they are independent, and therefore have a neutral effect. In an integrative design the functions and cues maintain a separate identity but are combined to enhance a product's beauty (e.g., a modular architecture). In integral, sculptural and organic design the functions lose their separateness and become part of the overall form of the product. The functions and cues are artistically represented within the overall form. A good example is an Apple Macintosh computer. The on/off button is flush with the surface of the monitor and artistically arranged to complement its surface colouring and texture.

Finally, each product is part of the aesthetic environment of the user. Each product contributes to the user's unique lifestyle. The products a product designer designs have to be appropriate for many people and each person uniquely. This is a challenging goal. One way of achieving the goal is to design products in such a way that they can be customised to match each user's lifestyle and fit in aesthetically with the other products that constitute the user's aesthetic environment. The user's environment is also the environment of other people. Aesthetic environments are social environments. So in developing products, the impact on other people must be considered as well.

It may not be feasible or practicable to provide all the user requirements in the product itself. In these situations, the product can be augmented with different forms of media and services. Different media can promote identification and the uniqueness of a user's lifestyle. Technology can be made visible by requiring customers to make or assemble their products or by allowing them to see their products being made. For example, Volkswagen has developed a transparent factory for Volkswagen Phaeton production at Dresden—the factory's walls are made almost entirely of glass. Customers and tourists can see cars being assembled.

Aesthetic Guidelines for Regenerative Development

Guidelines for a SMMC:

- Design products and manufacturing systems that are aesthetically pleasing and refrain from designing those that are aesthetically negative.
- Relate to stakeholders in ways that improve their aesthetic capability.

Guidelines for a designer:

- Be familiar with different design styles.
- Have a sound knowledge of environmental aesthetics.
- Be experienced at wandering in natural and urban landscapes.
- Study and appreciate as widely as possible humanity's natural and cultural heritage.
- Design products and manufacturing systems that are aesthetically pleasing. Refrain from designing products and environments that are aesthetically negative or neutral.
- Follow Alexander's design guidelines (Chapter 3).
- Design products to harmonise with users' aesthetic environments and lifestyles.

Guidelines for a designer who is a leader of a design team:

- Be an artistic leader. Develop the ability to inspire team members with visions of the future, inspire creativity, and help team members relax when they are exhausted.
- Ensure team members regularly engage nature by wandering.
- Provide resources to inspire heroism, creativity and catharsis.

Exercises and Problems

1. Briefly describe the meaning of the ideal state of re-creation. Give examples of institutions in the area where you live that contribute to this ideal.
2. Briefly describe environmental aesthetics.
3. Wander through an urban or natural landscape. Record the experience in your reflective journal.
4. Select one example of a consumer product that you consider to be a good design and one a bad design. Explain the basis for your judgements.

References

Ackoff, R.L., and Emery, F.E. 1972. *On purposeful systems*. London: Tavistock Publications.

Alexander, C. 2002. *The nature of order. Book 1: The phenomenon of life*. Berkeley, CA: Center for Environmental Structure.

Arieti, S. 1976. *Creativity: The magic synthesis*. New York: Basic Books.

Aspelund, K. 2010. *The design process*. 2nd ed. New York: Fairchild.

Berleant, A. 1997. *Living in the landscape: Toward an aesthetics of environment*. Lawrence: University Press of Kansas.

Bürdek, B.E. 2005. *Design: History, theory and practice of product design*. Basel: Birkhäuser.

Fenner, D.F.W. 2003. *Introducing aesthetics*. Westport, CT: Praeger.

Heyd, T. 2007. *Encountering nature: Toward an environmental culture*. Aldershot, UK: Ashgate.

Levinson, J. (ed.). 2003. *The Oxford handbook of aesthetics*. New York: Oxford University Press.

Lorand, R. 2000. *Aesthetic order: A philosophy of order, beauty, and art*. London: Routledge.

Norris, R.P., and Hamacher, D.W. 2010. Astronomical symbolism in Australian aboriginal rock art. *Rock Art Research*, 28(1), 99–106.

Singer, E.A., Jr. 1923. *On the contented life*. New York: Henry Holt and Co.

7

Social Context

An Ecological Perspective

Chapter 2 showed that an open system can be defined by four types of relations, three of which relate to its environment. The response capability of an open system is limited by the environment, especially the L_{22} relations. These are the most difficult to adapt to because an individual system often has little influence over them. Small to medium manufacturing companies (SMMCs), being open systems, need to have a good understanding of the L_{22} relations so they can survive and thrive. Fortunately, these kinds of relations have been studied and described by social scientists. We describe their work in this section.

Social scientists Emery and Trist classify environments according to the dynamic characteristics of the L_{22} relations, starting with static (unchanging) relations and moving to those that are complexly connected and highly dynamic (Emery 1974, Emery and Trist 1975).

The simplest type of environment is called a placid random environment. It is a simple, static environment with no connections between its parts, hence placid. The parts are distributed randomly and independently, hence random. The market example is perfect competition or an open price-takers market, which is the kind of market many jobbing manufacturers are in (Chapter 4). The appropriate planning mode is the tactic of doing one's best on a local basis.

The next environment is more complex because it has serial connection between its parts, but no feedback. It is called a placid clustered environment. It is placid because the parts do not change. It is clustered because some of the parts are connected together in clusters. The market example is imperfect competition (Chapter 4). Many SMMCs operate in this kind of market environment. Furthermore, the global hidden champions deliberately create this kind of market environment to prevent competition. The appropriate type of manufacturing is batch manufacture. A batch manufacturer can build up brand and company loyalty with a cluster of customers. A customer will not change brands unless the brand is not available or a price differential exists for a relatively long period of time between the brand the customer prefers and other brands available in the market. The clustering of parts means that some parts can act as signs for other parts. A company can use these signs to improve its knowledge of the market. Strategic planning is required, supported by tactical planning.

The next level of environment is called a disturbed reactive environment. "In this we simply postulate a placid clustered environment in which there is more than one system of the same kind, and hence the environment that is relevant to the survival of one is relevant to the survival of the other" (Emery 1974, p. 26). The resulting interactions create dynamics, but the background is stable. It is disturbed reactive because companies contribute to the dynamics (disturb it) through their market behaviour and also react to the induced dynamics. The market example is oligopoly (Chapter 4). Mass manufacturers typically operate in this kind of environment (in terms of market). Operational planning is required, embedded within strategic planning and supported by tactical planning.

The most complex environment within which adaptation is possible is the turbulent environment. The disturbed reactive environment results from interaction between similar systems. The turbulent environment is different because the dynamics arise from the environment either independently of the systems, e.g., floods and other natural disasters, or from social processes set off by the interacting systems. For companies, the market is no longer the most significant part of the environment; other parts, such as biological and sociological, must be adapted to. There is uncertainty about the environment in terms of which parts are likely to be affected by a company's actions and what the effects are likely to be. An individual company has little or no influence over the environment. A good example of turbulence is overfishing. Competitive strategies among fishermen based on an assumption of a static biological environment can, by overfishing, set off disastrous dynamic processes in the fish population, resulting in all the fishermen going out of business because there are no fish to catch.

> Turbulence weakens the very basis for purposeful adaptive behavior. Individual and organizational decisions become more difficult, because the environment is changing more rapidly and becoming more complex. The value basis for making decisions is itself thrown into question and becomes increasingly fragmented, creating new stresses throughout a social field. As a result, collective decisions become more difficult and consensus harder to achieve. Maladaptive effects of poor decisions are amplified as they interact with other environmental forces (the products of maladaptive responses in other parts of the extended field) to heighten turbulence, making future decisions and the discovery of adequate values more difficult. Turbulence, as a by-product of inadequately regulated social processes, contributes cumulatively to the complexity of system interactions and thereby to the difficulty of system regulation. Thus, the most crucial consequence of turbulence is that it makes adaptation to it ever more difficult. (Burns 1983, pp. 19–20)

Turbulent environments are extremely complex. It is not the complexity though that is the challenge; it is the increase in uncertainty. Turbulent environments are highly unpredictable. Actions often lead to counter-intuitive outcomes, adding to the uncertainty.

Due to the rich interconnections between all parts of the social environment, it is not possible to separate an individual system, e.g., a company, from the environment. Furthermore, an individual system cannot adapt to a turbulent environment on its own; other people and social systems are necessary too. For this reason, social scientists do not refer to individual systems in a turbulent environment, but to the whole interconnected set—the whole social field—which they call a social ecology. A turbulent world is an ecological world.

Emery and Trist's research led them to conclude that social turbulence occurs because of:

- Large organisations and sets of organisations whose actions deeply penetrate social fields and persist for long periods of time. The actions both initiate and amplify social processes, destabilising the social fields.

- Stronger linkages between people as makers and consumers and their other roles in life. These interdependencies have been intensified by the large size of companies.

- Widespread use of science to improve competitiveness. This increases the rate of change and creates stronger interdependencies between companies and their environments, due to the larger scale of operations.

- "The radical increase in the speed, scope, and capacity of intra-species communication.... The change that has taken place in intra-species communication is a greater mutation than if man had grown a second head. The consequences are a great increase in the information burden and radical reduction in the response time of the system—a reduction that is unaffected by distance" (Emery 1974, p. 30).

The types of environments are ideal types. The actual environment of a SMMC may be made up of more than one type. However, the turbulent environment has become more pervasive within nations and globally. Global SMMCs must develop organisational structures and processes to survive in the turbulent environments they will surely encounter.

Maladaptive Responses to Turbulence

All responses to social environments are based on values. Turbulence makes it very difficult to know what values are relevant and important. Maladaptive responses occur when people reduce the complexity by making simple choices of values. Effectively, it results in responses that limit the number of people that an individual person will cooperate with. Others are considered not relevant or are not tolerated. Whilst this makes adaptation easier—it reduces the information burden—it does not remove turbulence and more often intensifies it. Adaptive responses, on the other hand, attempt to identify emerging new values and embrace value diversity. They enlarge the number of people an individual person will cooperate with, and hence stabilise the turbulent social field.

The rise of the 19th-century industrial society destroyed the craft culture that preceded it and re-instituted ancient, bureaucratic forms of organisation—this is well documented in Mumford (1971). Emery argues that bureaucratic organisational structures not only prevent individual people from successfully adapting to turbulence, but also actually promote maladaptive responses. Emery and Trist (1975) have identified three maladaptive responses: segmentation, superficiality and dissociation.

Superficiality is "achieved by denying the reality of the deeper roots of humanity that bind social fields together and on a personal level the reality of their own psyche" (Emery and Trist 1975, p. 58). There are two common modes of superficiality. In countries controlled by autocratic governments, superficiality is expressed in terms of stereotypical behaviour and widespread use of violence to enforce compliance. In democratic countries it is expressed in terms of trivialisation. Anything and everything is pursued, so choice becomes irrelevant. The aim is to diversify choices as much as possible so the impact of any one choice has little effect. Real needs are ignored. Consider reality shows, for example. For many people these shows are as real as the real world they live in—What can be more superficial than that? People who cannot deny their human roots and express themselves uniquely are treated as outsiders because they will not conform to the (trivial) social norms.

Segmentation is based on prejudice. The individuals a person will cooperate with are limited to those the person knows and trusts. The social field is "transformed into a set of social fields each integrated in itself but poorly integrated with each other" (Emery and Trist 1975, p. 58). The disintegration into smaller social groups increases turbulence because it leads to less understanding between cultures and creates disadvantaged groups in society. In extreme cases segmentation can set off vortical processes at the boundaries between segments, leading to riots and wars.

Dissociation is an individual response to turbulence. It is a withdrawal from society due to fear of what involvement with others might lead to. For example, you might help someone in need in the street, being a good Samaritan, and then end up being robbed or sued by the person you helped. After that experience you decide not to help anyone in distress. Dissociation is more likely to occur in large cities due to the anonymity of people in a city. "The British seem more likely to retreat to their suburban castles and dissociate, on the grounds of 'I don't want to know', while the Americans and Australians defend their superficiality with 'So what?' or 'I couldn't care less'" (Emery and Trist 1975, pp. 65–66). Internet technology supports dissociation. People can retreat to their homes and be entertained by and purchase most, if not all, items they need through the Internet. They can have very little direct interaction with other people. Dissociation does not impact social fields directly, but indirectly by either segmentation or superficiality.

The three responses are mutually reinforcing. They are the result of a failure to recognise the complexity of turbulent social environments and are not capable of removing turbulence. Quality of life is reduced, though the standard of living may be high. Worse still, there is a high probability that vortical processes will be set off at segment boundaries, resulting in violence. This can be clearly seen in the "Arab spring" revolts in the Middle East.

Adaptive Responses to Turbulence

The only effective way to remove turbulence is through values that have an "over-riding significance for members of the field. Values have always arisen as the human response to persisting areas of relevant uncertainty" (Emery and Trist 1975, p. 68). The values must be long-term values sought by all members of the field; in other words, they are ideals. We have now come full circle: from arguing how one ought to live a good life by pursuing ideals to stating that ideals are the only way that human societies can successfully adapt to turbulent social fields.

Ideal pursuit requires cooperation between people. So the question now is: How can individuals and SMMCs cooperate to remove turbulence in social fields? Emery's view is that it is best to develop individual competence in pursuing ideals by developing non-bureaucratic forms of organisation based on redundancy of function rather than redundancy of parts. People will then be able to change companies to support their values, and they will be able to use their competence in other aspects of life, reducing turbulence. Section IV explains how to set up non-bureaucratic manufacturing companies and design teams. SMMCs also need a process through which they can explore the social fields to find shifts in values and to incorporate the people and social systems with these values as stakeholders. One such process is interactive planning.

Interactive planning is a proactive approach to remove turbulence. It is a process containing five phases. The first phase, mess formulation, includes a search for values in the social field and how these may affect the SMMC. The search for values can be conducted using a search conference (Emery and Devane 1999). The second phase, idealised designing, is aimed at re-designing the SMMC so all stakeholders can pursue ideals. The design will include a statement of purpose (Chapter 5).

The statement of purpose is expressed in terms of how a SMMC intends to relate to its stakeholders; it defines the required norm. Senior management of the SMMC is accountable for regulating the relationships (Vickers 1995). Regulation is a process, not a goal. There are two regulative strategies: norm holding and threshold avoidance. Norm holding is aimed at maintaining the relationships within an acceptable level of variance. Threshold avoidance is aimed at preventing one or more relationships from reaching a critical threshold value at which the relationship will break down. Norm holding is to be preferred, as it lowers risks and leads to more stable relationships.

The decisions managers make to regulate the relationships are best described as judgements (Vickers 1995). A judgement is based on an appreciation of all the stakeholders' values in the social field, the interrelationships between the values, the change in values over time, and the relevant facts (constraints and requirements). Managerial judgements are the business equivalent of legal judgements. Managerial training in the art of judgement should follow a process similar to that by which legal judges are educated and trained.

Trist argues that all levels of society need to be considered to remove turbulence. There are two basic strategies for achieving cooperation between companies and other social groups:

1. Gain control of all elements in the environment (or a large portion of it) and impose a central value system. This is an approach used by governments to control exploitation of resources and pollution. The government sets the ground rules by law and through enforcement ensures these are obeyed.

2. Knit together voluntarily the autonomous parts of the environment to form a network that permits concerted action.

In the first strategy the value system is set by mandated means. The economic example is macro-regulation. Unfortunately, many turbulent processes now cross national boundaries, and the first approach has very limited value, e.g., regulation of the banking system, control of pollution that crosses national borders, and regulation of water usage and quality for rivers that flow through several countries. Furthermore, Trist argues macro-regulation is not sufficient even within a nation. So macro-regulation, whilst it may be essential in some situations, is insufficient in all. The second strategy is far more important.

In the second strategy a combination of voluntary (incentive systems) and mandated (legal) means are used. The mandated means are required to enforce compliance when members of the network break the norms and agreements. Within nations networks need to be created at the community and regional levels. Global networks need to be created to remove worldwide turbulence between nations and global social groups.

The management competencies needed for this strategy are creativity, leadership, entrepreneurship, judgement and consensus building skills.

It should be noted that large size does not guarantee a company will be successful in a turbulent environment. Drucker (1980) argues that the multi-national corporation needs to change its structure to become a transnational federation of autonomous units.

Social Guidelines for Regenerative Development

The following guidelines will help SMMCs adapt to turbulent social fields:

- Create non-bureaucratic forms of organisational structures and promote democratisation of the workplace (Section IV).
- Take an active approach to identifying emerging values of relevance to operations.
- Take an active approach in initiating, creating and supporting cooperative networks. We recommend interactive planning, as it was specifically designed for this purpose.
- Promote and train managers to be creative, entrepreneurial leaders who can make judgements guided by ideals.

We conclude the chapter by describing some social trends that are likely to have significant impact on global SMMCs for the next 30 years. These trends were identified by leading researchers in the 1970s (Emery and Trist 1975, Drucker 1980) and some are only now becoming widespread. We have chosen the original sources to indicate the power of early identification of social changes.

- From succession to simultaneity. The traditional (successive) mode of working is school, work and retire. The new mode of working is to undertake learning, working and leisure simultaneously. Hence people can be expected to work longer and retire later. This trend has taken over 30 years to become significant. SMMCs should design work roles and organisational structures to support simultaneous learning, working and leisure and a more heterogeneous work force (from young to old).
- From stasis to mobility. People are becoming more mobile and are changing jobs much more frequently than in the past. Staff turnover within a global SMMC can be reduced by providing opportunities for employees to grow and achieve their desired goals. The global nature of the business means that employees can be given the opportunity to travel as part of their job.

- From goods to services. Once a certain standard of living has been achieved, people focus more on services rather than on goods. SMMCs should focus less on products and more on providing services around products or look for opportunities to provide services only. Back-end service costs can be reduced by delivering these services through emerging or developing economies where labour costs are low. India is a good example.
- Changing demographics. The populations are aging in developed economies, whereas developing economies have very young populations. In Saudi Arabia, for example, 70% of the population is under the age of 30. Developing economies struggle to create jobs to employ the young people. There are opportunities for global SMMCs to locate in these countries and provide meaningful employment for the young people. In the developed economies there will be a shortage of young people entering the workforce due to low birth rates, although the shortage will be partially compensated for by the longer working life of older people.
- Older people are basically consumers—they spend and do not need to save. There are market opportunities for SMMCs to provide products and services to meet the needs of older people.

Exercises and Problems

1. Discuss the relevance of turbulent social environments to SMMCs.
2. How can a SMMC successfully adapt to a turbulent environment?
3. A local SMMC is a contract manufacturer for a large multinational company. The CEO of the SMMC argues that he does not have to worry about turbulent environments because he is protected by the MNC. Is he correct? Explain your answer.
4. What are the population demographics for the country within which you live? What is the most likely scenario for the change in demographics over the next 30 years?

References

Burns, T. 1983. *The active adaptive systems paradigm: Underlying theory.* S³ Paper 83–02. Philadelphia: Social Systems Sciences Department, Wharton School, University of Pennsylvania.

Drucker, P. 1980. *Managing in turbulent times*. London: Pan Books.

Emery, F.E. 1974. *Futures we're in*. Canberra: Centre for Continuing Education, Australian National University.

Emery, F.E., and Trist, E.L. 1975. *Towards a social ecology*. New York: Plenum.

Emery, M., and Devane, T. 1999. *Search conference*. San Francisco: Berrett-Koehler Communications.

Mumford, L. 1971. *The pentagon of power; the myth of the machine*. London: Secker and Warburg.

Vickers, G. 1995. *The art of judgment*. London: Sage. (First published in 1965.)

8

Technological Context

Innovation

There is a difference between an invention and an innovation. An *invention* is a new idea, concept, product or process or new modification to existing products and processes. It is something that is new and technically feasible. Often the new idea is legally protected by law through a patent or copyright. An *innovation* is an invention that has been commercially exploited. It is technically and commercially feasible and socially desirable.

Innovation is a central feature of a modern economy. The main difference between innovation today and that carried out prior to the 1900s is that research and development (R&D) has become a specialised professional activity: highly skilled scientists and technologists work full-time in both private and government organisations carrying out R&D. This has had two main effects. First, larger projects can be undertaken, e.g., space exploration and nuclear power stations. Second, the rate of technological change is faster.

Small companies can innovate. They can even compete against companies in industries with very high R&D minimum threshold expenditures, e.g., disk drives, steel mills and even satellites. They do this by not competing for the high-quality end of these markets, which require the very high R&D expenditures. Instead, they invade the low-quality end, which does not require large expenditures. As they use more cost-effective technology than the high-end producers, they are able to make products at much lower costs and at the same time achieve good profits. These profits are fed back to upgrade the technology. This process continues until they have developed their technologies to compete directly in the high end of the market. Unfortunately, when they get there, the prices have been reduced by competition and the profit margins are much lower than when they were in the low end of the market. This kind of innovation process is called disruptive innovation (Christensen 1997). We do not recommend it to small to medium manufacturing companies (SMMCs) because the end result is always the same: reducing profit margins as the disruptive companies improve the quality of their products, and then, finally, the danger that a small company somewhere will take your market away when you reach the high-quality end of the market.

This happened to the United States when Japan invaded its markets in the 1970s, and is happening to Japan today as Korea invades its markets.

We now look at the innovation strategies that are available to companies. The groundbreaking work in this area was carried out by Freeman in the 1980s and has been corroborated by recent research studies. The strategies are traditional, dependent, imitative, defensive, offensive and opportunistic (Freeman 1982, Martin 1994). These strategies are pure or ideal types. In practice, companies may use a mix of strategies.

> *Traditional (me evolve slowly strategy)*: This strategy can be characterised as one of minimum innovation. A traditional company changes its products very little. It lacks scientific and technological capability to initiate significant product changes, but is able to produce minor design changes. This is its main strength. Traditional technology is often based on craft skills that evolve slowly. Product demand can be high when the craft skills come from a long tradition, e.g., handicrafts. These companies can survive even in highly industrialised economies. On the other hand, many have not survived because they have not been able to adapt to changes in technology in their industrial sectors.

> *Dependent (you to me strategy)*: This strategy is innovation from without. A dependent company is a satellite of a larger parent company, either as a subsidiary, department, sub-contractor or a licensee. Technical expertise is provided by the parent company and the customers. A dependent company has no initiative for product design. Design changes are made only on the request of customers or the parent company. It has no R&D capability. The satellite company can "often enjoy good profits for considerable periods, because of low overheads, entrepreneurial skill, specialized craft knowledge or other peculiar local advantages. Even if they are 'squeezed' pretty hard by their customers, they may prefer to endure long periods of low profitability rather than be taken over completely" (Freeman 1982, p. 182).

> *Imitative (me too strategy)*: This strategy is incremental innovation (from within). An imitative company copies the innovations of other companies. If the time lag is short between the copy and the innovation, then the imitative company may need to get a licence or buy the technological knowledge from the innovating company. If the time lag is long, then a licence may not be necessary. Imitative companies often survive because of market protection or special local advantages, such as lower energy or labour costs than their competitors. If they have no special advantages, then they must have lower production or operating costs than their competitors in order to compete; i.e., they must be more efficient. Hence an imitative company has a strong technical capability in production engineering and design.

It also has a good scientific and technical information system in order to keep up with technical developments in its industry. An imitative company may patent its innovations. However, its income from licensing will always be less than its expenditures for licences and technical knowledge.

Defensive (me second strategy): This strategy can be characterised as a mixture of incremental and radical innovation. A defensive company does not aim at being the first in developing a new radical innovation, but neither does it want to be left behind. "They may not wish to incur the heavy risks of being first to innovate and may imagine that they can profit from the mistakes of the early innovators and from their opening up the market. Alternatively, the 'defensive' innovator may lack the capacity for the more original types of innovation, and in particular the links to fundamental research. Or they may have particular skills in production engineering and marketing" (Freeman 1982, p. 176).

A defensive company needs to come up with its own designs, which are as good as or better than those of the early innovators, if it wants to capture a large share of the market. Therefore it must have an R&D function. The amount spent on R&D by a defensive innovator may be larger than that by an offensive innovator because the defensive innovator may have to move more rapidly.

A defensive innovator needs to be reasonably strong in providing technical advice and the education and training of its customers. A combination of technical assistance and product differentiation is a powerful way of obtaining market share. Long-range planning is also important.

Offensive (me first strategy): This strategy can be characterised as radical innovation: its objective is technical and market leadership. The offensive innovator needs to be strong in education and training of its customers and its own personnel in order to "socialize" its innovations. It also needs to be strong in long-range planning, production and marketing. Patent protection is very important for an offensive innovator, as a monopoly on profits is needed to recover the high investment in R&D and the failures.

Opportunistic (me only strategy): This strategy can be characterised as niche innovation: its objective is market leadership. Market leadership is obtained by finding gaps in the current technology-market mix and exploiting them. This kind of innovation does not require a research function. It does require development capability and imaginative entrepreneurship. There are niche market opportunities for:

- Newly innovated products and processes (within the current technical know-how)

- Ingenious adaptations of mature products and processes
- Application of existing products and processes to new markets
- Ingenious assemblies of mature or new products

Any new product or process has a product life cycle. In general, several or many companies will produce the activities in the life cycle. The type of innovation they can undertake depends on which activities they produce: R&D, design, manufacture, market, maintain and operate (use). The relationship between the activities and innovation objectives and strategies is shown in Table 8.1. As an example, consider the first activity, "operate only", and assume the product is a commercial aircraft. An airline using the aircraft cannot experiment with new ways of operating it. It is limited to operating the aircraft as effectively as possible within the technological and legal constraints. It can attain a distinctive competence in its use, e.g., Singapore Airlines. The airline is dependent on the aircraft supplier for innovation in aircraft technology (dependent strategy). It may be able to evolve better ways to provide cabin service to customers, but given the constraints, these will evolve slowly over time (traditional strategy). Note that we are only discussing the use of the aircraft. An airline can innovate radically by providing a new service concept, e.g., a budget airline service.

The best innovation strategies for SMMCs are traditional, imitative and opportunistic. These strategies require SMMCs to obtain their technology from external sources. External sources of technology include:

- Technological and scientific infrastructure, e.g., universities, scientific research centres, etc.
- Indigenous cultures
- Literature, including patents

TABLE 8.1

Innovation Objectives and Strategies

Activity	Innovation Objective	Appropriate Innovation Strategies
Operate only	Distinctive competence	Traditional, dependent
Operate and maintain	Distinctive competence	Traditional, dependent, imitative
Market only	Distinctive competence	Traditional, dependent, imitative, opportunistic
Manufacture, market and maintain	Market leader	Traditional, dependent, imitative, opportunistic
Design, manufacture, market and maintain	Market leader	Imitative, opportunistic
R&D, design, manufacture, market and maintain	Technical or market leader	Defensive, offensive

- Leading-edge customers
- Leading-edge suppliers
- Competitors
- Recruitment of expert personnel who have the desired technical knowledge and skills
- Acquisition of, equity investments in, or partnerships with companies that have the desired technologies

Indigenous cultures are often overlooked as a source of technological information. Yet, these cultures have long histories of living in harmony with the environment. More importantly, they have integrated the aesthetic and moral aspects with the technical (Heyd 2007). We can learn a lot from them. Indigenous cultures are a useful source of technical knowledge in health, agriculture and environmental ecology. One of the best examples of successful indigenous local knowledge is the water temples in Bali. The temple priests coordinate the activities of many rice farmers, resulting in proper water and irrigation management and pest control. This traditional practice has worked effectively for over 1,000 years. But in 1967 Indonesia initiated a green revolution to increase rice production. The government forced farmers to stop using the water temples and to use high-yield rice seeds, fertilizers and pesticides instead. Farmers were encouraged to grow three crops per year. By 1974 the results of the green revolution were clear: chaos in water scheduling, a large increase in pests and a subsequent fall in rice production. Scientific studies by Dr. Lansing proved the effectiveness of the water temples for water management and pest control. His studies led to the Indonesian government relenting in the 1980s and allowing farmers to return to their traditional method of rice cultivation. The water temples can be seen in Bali today.

SMMCs create knowledge during the product development process. The organisational and design team designs described in Section IV actively involve stakeholders in the design process. Therefore they also help stakeholders gain knowledge.

Limits to Business Development

In a little-known book, Larsson (2004) presents a strong argument that there are real limits to business development. He says that new businesses only improve the automation of processes that people have been doing for hundreds of years. The limits of the automation are zero time and zero cost. He presents a series of case studies in different areas, such as information and transportation, to defend his thesis. The best example is information.

We have now reached a stage where information can be transmitted almost instantaneously (zero time) and at very little cost.

1. Companies and organizations are in many areas of their businesses coming closer and closer to definite limits of business development. Nothing can be done in less than no time and at less than no cost and in many cases companies are closing in on these limits....

2. Automation of time and cost reduction is going on in all areas of society at the same time. There are no obvious industries that are likely to need more resources in the future....

3. Through automation process steps are integrated and companies are becoming less unique. It becomes more and more difficult to develop unique and sustainable competitive advantages. Competitors become more and more equal and profit margins decrease. (Larsson 2004, p. 22)

Larsson is referring to manufacturing industries. His arguments do not hold for the extractive industries, which will require more resources in the future to extract the lower-quality materials remaining (Chapter 9).

What are the implications for SMMCs? First, as lead times are compressed and costs lowered, the cost of producing additional increments of reduction increases. A point is reached where it is not economically worthwhile to compress time or reduce costs further.

Second, if SMMCs adopt the practices in this book, then they will be operating at an appropriate level of time compression (minimum feasible lead time) and low cost.

Third, when the minimum time and cost targets are reached, companies will have to find other ways to generate competitive advantage. This can be done through developing *secondary qualities* of products and manufacturing systems. These secondary qualities are values other than time and cost. They are "qualities that are related to the way that a product is produced, the way that the producing or distributing company is owned, its size or the policies of employment that are in use by a company" (Larsson 2004, p. 174). In other words, they relate to the moral, environmental, political and aesthetic aspects of life.

Regardless of which industry a SMMC is in, the limit of time compression and cost reduction will be reached. At that point, the SMMC will have to turn to the secondary qualities for development. For reasons outlined in the next chapter, it is better to start now rather than wait, especially on qualities relating to the environment. Environmental ecological principles can be applied to industrial supply chains. There are three major technical principles for designing industrial supply chains as ecological networks (Edwards 2010):

- Cradle-to-cradle manufacturing: Products are designed so that they can be recycled into new products at the end of their life (re-purposing).

- Waste equals food: The waste energy and materials from one company are the inputs to another. "In nature, there is no waste: nutrients support metabolic processes. The compost from decayed plants, for example, provides food for living plants. In the manufacturing sector, industrial ecology sites such as Kalundborg, Denmark, mimic these processes. Excess heat and material from a power plant, a water treatment plant, a refinery and a biotech company are reused" (Edwards 2010, p. 53).

- Stay within current solar income. "Renewable resources must be consumed at rates below those of natural replenishment.... Non-renewable resources must be consumed at declining rates (with rates of decline at least equaling rates of depletion), and recycled wherever possible" (Heinberg 2011, p. 247).

Global Risk Management

Global SMMCs face many risks. Factories can be closed because of fire, flooding, earthquakes, tornadoes, hurricanes, riots, etc. A business continuity plan (BCP) identifies the anticipated risks that can temporarily stop the operations of a SMMC or disrupt its supply chain, and specifies the response and response time for each threat.

The steps for developing and implementing a BCP are:

1. Create basic emergency procedures. You must have basic emergency procedures in place first. These include evacuation procedures, how to deal with workplace injuries, etc.

2. Profile the supply chain. The profile must include suppliers as well as assets owned by the SMMC. At the very least, you should consider immediate suppliers and those once removed. Take particular note of sole source suppliers. If these stop supplying, then the entire supply chain is stopped. Several Japanese companies found this out the hard way during the flooding of Bangkok, Thailand, in 2011. They had to close down their entire manufacturing system for some products because critical parts and materials were not available— they were provided by sole source factories in Bangkok. Sole source suppliers are high risk.

3. Identify and classify the risks. Some of the major risks for small businesses are fires, floods, earthquakes, hurricanes/typhoons, riots, terrorism, utilities failure, theft, strikes, infectious diseases and legal/regulatory action. Risks can be classified into three categories: high, medium and low. High-risk characteristics include

loss of life or complete closure of the supply chain for a significant period of time, e.g., 1–2 months. Medium-risk characteristics include non-life-threatening injuries or partial closure of the supply chain or complete closure for an intermediate period of time, e.g., 1–2 weeks. Low risk includes no injuries or very short cessation of operations in one part of the supply chain, e.g., less than 1 week.

4. Assign ownership for risks and develop contingency plans. The plans should define the response procedures, permissible response times and estimate how long it will take to get operations up and running again for each type of risk.

5. Write the BCP. Items that should be included in the BCP are emergency personnel names and contact details, emergency contact numbers for staff, external emergency services contact numbers, list of key customers and suppliers and contact details, identification of risks and critical business processes, and contingency plans.

6. Test the plan. It is good practice to test the plan by simulating disasters and emergencies.

7. Approve and disseminate the plan.

8. Update the plan regularly.

Technical Guidelines for Regenerative Development

The following technical guidelines will help SMMCs develop products and manufacturing systems for regenerative development:

- Use traditional, imitative and opportunistic innovation strategies.
- Seek technology from all sources, including indigenous cultures.
- Promote and support indigenous knowledge as well as that developed by modern science.
- Develop the technical capabilities of stakeholders through lifelong learning.
- Adopt the principles and practices outlined in this book in order to reduce lead time and costs.
- Focus on secondary qualities to develop products and manufacturing systems. Implement the three technical principles for designing industrial supply chains as ecological networks: cradle-to-cradle manufacturing, waste equals food, and staying within solar income. This includes adopting the following practices:
  - Energy reduction
  - Waste reduction

- Material use reduction
- Switching from non-renewable to renewable energy resources
- Designing products for long lives or re-purposing
- Develop and implement a BCP.

Exercises and Problems

1. Describe the different innovation strategies.
2. Why are there limits to business development?
3. Identify the major risks for operating a factory in (a) Bangkok, Thailand, (b) New Delhi, India, and (c) Singapore. (You will have to do some research to find out what these are.)

References

Christensen, C.M. 1997. *The innovator's dilemma: When new technologies cause great firms to fail.* Boston: Harvard Business School Press.

Edwards, A.R. 2010. *Thriving beyond sustainability.* Gabriola Island, BC: New Society Publishers.

Freeman, C. 1982. *The economics of industrial innovation.* London: Frances Pinter.

Heinberg, R. 2011. *The end of growth: Adapting to the new reality.* Gabriola Island, BC: New Society Publishers.

Heyd, T. 2007. *Encountering nature: Toward an environmental culture.* Aldershot, UK: Ashgate.

Larsson, M. 2004. *The limits of business development and economic growth: Why business will need to invest less in the future.* Basingstoke: Palgrave Macmillan.

Martin, M.J.C. 1994. *Managing technological innovation and entrepreneurship in technology based firms.* New York: John Wiley.

9

Environmental Context

Environmental Morality

One of the major issues confronting environmentalists is how to encourage people to feel responsible for the environment. Three philosophical approaches have been proposed. The first is an economic approach. Proponents of this approach argue that the environment should have an economic value. By setting the market price high enough, people will be encouraged to take care of the environment. This approach is immoral, as Churchman (1979) and others have shown. So it cannot be used as a moral justification.

The second approach is to argue that damaging the environment affects other people either directly or indirectly. As people are significant others, we should therefore take care of the environment. This approach has moral justification. However, it ignores the fact that indigenous small communities, some Eastern religions and philosophies, and environmentalists have a love of and feel a duty of care to the environment itself. This leads to the third approach. Here the concept of significant other is extended to include non-human entities; these can be plants, animals, landscapes and features of landscapes. The moral justification now is that one should take care of the environment because it is a significant other (Heyd 2007). This argument is complementary to Alexander's view that everything has life (Chapter 2).

Chapter 5 noted that morality is affirmed and reaffirmed by our relationships with other people. The extended morality proposed here means that we need to affirm and reaffirm our relationships with the environment. Many people live in urban environments and have little direct contact with natural environments (apart from the sky and perhaps the sea). Without frequent contact with natural environments they will not develop a morality toward it. Therefore it is necessary for people to regularly visit natural environments. The best way to appreciate them is by wandering (Heyd 2007).

Environmental morality should be part of our everyday life at work and at home (Heyd 2007). Unfortunately, the environment is often treated as a dumping ground for rubbish and other unwanted items. Even with good intentions our disposal of rubbish can lead to unforeseen environmental damage. For example, many plastic products reach the sea. Ocean currents move the waste to areas (called gyres) where the currents slowly circulate. Once the waste reaches a gyre it is trapped—a gyre is a gigantic waste disposal sink. Plastic degrades slowly and breaks down into very small particles, which are ingested by fish and birds, killing them. The North Pacific gyre is the most infamous and is euphemistically referred to as the Great Pacific Garbage Patch. Designers and small to medium manufacturing companies (SMMCs) can help reduce waste pollution by designing products to be re-used, recycled, re-manufactured, or biodegradable.

The disposal of unwanted items is one kind of pollution. There is another kind where the polluter desires the polluting action. For example, noise pollution occurs when one person plays loud music, which he enjoys, but which annoys other people. It can also occur in light pollution, where one person lights up an area that other people prefer to have dark or interferes with noc-turnal animals. Designers and SMMCs can help reduce this kind of pollution by designing products that allow people to enjoy the sight, sound, etc., without disturbing other people or animals, such as headphones for music.

Governments try to prevent pollution through:

- Legislation and enforcement
- Taxing pollution (Heinberg 2011, Jackson 2011).
- Encouraging industry to implement waste equals food practices (Chapter 8)
- Providing green spaces, so people can appreciate nature
- Public education programs

Limits to Growth

In 1972 one book stirred up major outrage amongst economists, scientists, politicians and industrialists. The title of the book was *Limits to Growth* (Meadows et al. 1972). The book published the results derived from a physico-socio-economic simulation model of the world developed by a team of researchers at Massachusetts Institute of Technology (MIT). The controversy occurred because the model showed there are limits to resource usage and population growth. The limits are due to the finite capacity of earth's resources and its finite ability to absorb pollution. At the time of publication these views were considered heresy—hence the outrage. In this section, we explain what the model is, what it does, what its results portend for our future and answer the critics.

The model, *World3*, is based on a system dynamic simulation language invented by Jay Forrestor in the mid-1950s when he was a professor at MIT. The simulation model is a closed system model: it models a system *and* its environment; in other words, it models an ecology.

One of the earliest models was of a supply chain (Forrestor 1961). The modelling showed how small increases in customer demand at the end of a supply chain led to large swings in orders at the beginning of it. These swings were produced by the inherent structure of the supply chain and were not due to external factors. *This is the most important function of system dynamic modelling: to demonstrate that the dynamic behaviour of systems is due to their inherent structure. Therefore if you want to change the dynamic behaviour of a system, you have to change its structure.*

The simulation language models flows and inventories (storage points). Equations define the rate of flows into and out of inventories. Delays can be entered into the model as constants or equations. The model can represent physical entities, such as material and energy flows, and non-physical entities, such as information flows. The maximum capacities of inventories and flows are entered as limits. Some inventories will be sources of materials and energy, some will be endpoints (sinks) and the rest are in-process inventories.

The models are run on computers using a computer simulation language called *DYNAMO*. A model run is initiated by defining a set of starting conditions. The variables that define the starting conditions can have constant values during the simulation or can change in value, with the changes represented by equations.

System dynamic models do not predict specific outcomes. What they do is show the typical dynamic behaviour patterns of systems. These patterns derive solely from the structure of the systems. When the dynamic behaviour patterns are not acceptable, then the structure of the system has to be changed. This may involve changing policies, the links between the parts of the system, the flows or the delays. The changes are made to the model and it is run again and again, until acceptable or desirable patterns are achieved. In other words, systems dynamic models are used for scenario planning.

The sources for World3 are planetary sources for non-renewable and renewal energy and materials. These are inputs to an economic system that converts them to useful products and wastes and pollution, which end up in planetary sinks. Non-renewable sources clearly have finite limits; these are included in the model. Similarly, the earth has a finite capacity to absorb wastes and pollution. Renewal resources can also deplete if the rate of use is higher than the rate of replenishment. The MIT scientists state that they

> focused principally on the planet's physical limits, in the form of depletable natural resources and the finite capacity of the Earth to absorb emissions from industry and agriculture. In every realistic scenario

we found that these limits force an end to physical growth in World3 sometime during the twenty-first century.

Our analysis did not foresee abrupt limits—absent one day, totally binding the next. In our scenarios the expansion of population and physical capital gradually forces humanity to divert more and more capital to cope with the problems arising from a combination of constraints. Eventually, so much capital is diverted to solving these problems that it becomes impossible to sustain further growth in industrial output. When industry declines, society can no longer sustain greater and greater output in the other economic sectors: food, services, and other consumption. When these sectors quit growing, population growth also ceases.

The end to growth may take many forms. It can occur as a collapse: an uncontrolled decline in both population and human welfare. The scenarios of World3 portray such a collapse from a variety of causes. The end to growth can also occur as a smooth adaptation of the human footprint to the carrying capacity of the globe. By specifying major changes in current policies we can cause World3 to generate scenarios with an orderly end to growth followed by a long period of relatively high human welfare. (Meadows et al. 2004, pp. x, xi)

All scenarios result in an end to growth. The bad news is that most of the scenarios generate collapse. There are only a few scenarios that produce an orderly (controlled) end to growth.

World3 results are supported in detailed studies by some economists and energy/resource experts. Consider, for example, non-renewable resources. The extraction rates for non-renewable resources typically reach a maximum at some point and then decline. This phenomenon was first shown for oil (Deffeyes 2001). "The phrase 'Peak Oil' is often misunderstood to refer to the total exhaustion of petroleum reserves—*running out*. In fact it just signifies the period when the production of oil achieves its maximum rate before beginning its inevitable decline" (Heinberg 2011, p. 107). The reason for the peak is that exploration and extraction begins with the most easily accessible and largest regions (larger ones are easier to find). By corollary, the remaining regions are harder to find, smaller, and the quality is lower. Therefore it costs more to find and develop them. "The trends in the oil industry are clear and undisputed: exploration and production are becoming more costly, and are entailing more environmental risks" (Heinberg 2011, p. 110).

The same process is at work for other finite resources. Once a peak has been reached, more and more capital is required to extract the resources. Many resources are past their peak and others are expected to reach their peaks within the next 20 years. *Resource limit peaks for non-renewable resources are predicted by World3.*

One major criticism of World3 is that it is not an accurate representation of what is happening. There are two replies. The first is that it is the authors of World3 who acknowledge inaccuracies in the model. They carried out sensitivity studies to determine the impact of uncertainties in the model's parameters. These had no effect on the ultimate result—an end to growth, nor on the resulting dynamic patterns—collapse or orderly end to growth. Second, show us a better model. As far as we are aware, there is none.

A second criticism states that limits on current resources do not matter as we will invent new technologies that can substitute for them. It is assumed there is no limit to finding and implementing substitutive technologies. Meadows et al. (2004), Heinberg (2011) and Jackson (2011) clearly show that this argument is false. As an example, let us consider a replacement energy source for oil. Oil is so widely used because it is an excellent source of energy: it can be extracted relatively cheaply, it is very portable and it delivers a lot of energy per unit vol/wt (Heinberg 2011). The selection of an alternative energy source depends on the ratio of the amount of energy a source delivers to that required to produce it—energy return on energy invested (EROEI). It is estimated that the EROEI for oil is 100:1. It is also estimated that our modern industrial society requires a minimum EROEI of 10 to 5:1. Now you can see why oil is used. There is a tremendous energy surplus (10–20 times).

There is no known substitute fuel now or in the near future that has an EROEI anywhere near that of oil. Bio-fuels have EOREIs below 5:1. Wind power is estimated to have an EROEI range from 25 to 15:1. This is higher than the minimum requirement for an industrial society, but wind power requires large areas of land and is not reliable. It is interesting to note that nuclear power is on the borderline of the upper minimum requirement limit. Here is Heinberg's conclusion (2011, p. 117):

> In 2009, Post Carbon Institute and the International Forum on Globalization undertook a joint study to analyze 18 energy sources (from oil to tidal power) using 10 criteria (scalability, renewability, energy density, energy returned on energy invested, and so on)…. It was, to my knowledge, the first time so many energy resources had been examined using so many essential criteria. Our conclusion was that there is no credible scenario in which alternative energy sources can entirely make up for fossil fuels as the latter deplete. The overwhelming likelihood is that, by 2100, global society will have *less* energy available for economic purposes, not more.

Nurturing the Environment

It is possible to live within the earth's finite limits, but humans will have to change the way they live. Large-scale, resource-intensive, polluting networks and systems need to be replaced by small communities supporting local ecologies. Environmentalists refer to these as *community ecological systems* (CESs). A CES is a small geographically defined unit, such as a town or urban unit (for a city), that is self-sustaining as far as is practicable. Communities are the primary social unit for seeking ideals. They combine production (the economy), politics, morality and knowledge, and aesthetics (Churchman 1979).

Each CES will be supported by a community ecological laboratory (CEL). "The CEL would be a multi-function hub consisting of a number of independent organizations and businesses dedicated to helping people impacted by hard times, and to providing an armature around which a new economy could be woven. It will offer a variety of services, as well as opportunities for self-improvement, learning, enterprise incubation, and community involvement" (Heinberg 2011, p. 277).

CESs will abide by the following One Planet Living guiding principles (Edwards 2010, pp. 84–85):

1. **Zero carbon:** Our climate is changing because of human-induced build up of CO_2 in the atmosphere.
2. **Zero waste:** Waste from discarded products and packaging creates disposal problems and squanders valuable resources.
3. **Sustainable transport:** Travel by car and airplane is contributing to climate change, air and noise pollution, and congestion.
4. **Local and sustainable materials:** Destructive resources exploitation (e.g., in construction and manufacturing) increases environmental damage and reduces benefits to [the] local community.
5. **Local and sustainable food:** Industrial agriculture produces food of uncertain quality, harms local ecosystems, and may have high transport impacts.
6. **Sustainable water:** Local supplies of freshwater are often insufficient to meet human needs, due to pollution, disruption of hydrological cycles, and depletion.
7. **Natural habitats and wildlife:** Loss of biodiversity [is] due to development in natural areas and over-exploitation of natural resources.
8. **Culture and heritage:** Local cultural heritage is being lost throughout the world due to globalisation, resulting in loss of local identity and knowledge.
9. **Equity and fair trade:** Some in the industrialised world live in relative poverty, while many in the developing world cannot meet their basic needs from what they produce or sell.
10. **Health and happiness:** Rising wealth and greater health and happiness increasingly diverge, raising questions about the true basis of well-being and contentment.

Environmental Guidelines for Regenerative Development

SMMCs collectively account for most economic activity in an economy; e.g., they account for 70% of GDP in Singapore. They also employ most of the working population. So, while individually each SMMC is small, collectively SMMCs are large. SMMCs could use their collective economic and political power to ensure an orderly transition to a zero-growth, thriving, world economy. The following guidelines will help a SMMC support regenerative development.

- Take responsibility for improving the environmental morality of your stakeholders.
- Implement the three technical principles for designing industrial supply chains as ecological networks: cradle-to-cradle manufacturing, waste equals food, and staying within solar income (Chapter 8).
- Initiate and support CESs.
- Invest in projects that will improve the ecology (adapted from Jackson 2011), such as:
  - Retrofitting buildings with energy and carbon-saving measures.
  - Renewable energy technologies.
  - Redesigning utility networks. Large-scale networks should be broken up into local networks to support CESs.
  - Designing sustainable transportation systems to support CESs through local, manual, private transport, e.g., walking and cycling, and public transportation between CESs.
  - Designing public green spaces to support CESs.
  - Maintaining and protecting ecosystems.
- Support initiatives to promote and implement regenerative policies and practices, e.g., Richard Branson's initiatives the Carbon War Room and the B team (*The Economist* 2012). Edwards (2010) provides a long list of environmental organisations that a SMMC could support and work with.

The Singapore government has a number of initiatives to improve the environment. One highly successful urban rehabilitation project at Bishan Park won World Architecture Festival Landscape of the Year for 2012. The project converted a wide, very ugly, concrete industrial drain into a flowing river integrated into a park. Figure 9.1 shows the "drain" as it is now. The water level is low, but during the monsoon season the level can rise 1–2 metres. The river twists and turns and has rocks and plants to reduce the water speed during peak flow.

FIGURE 9.1
Rehabilitated industrial drain at Bishan Park, Singapore.

Exercises and Problems

1. Explain what environmental morality is.
2. Describe a community ecological system.
3. List the green spaces that are available to you locally.
4. Make a list of changes you can make in your lifestyle so that it conforms more closely with the One Planet Living guiding principles.

References

Churchman, C.W. 1979. *The systems approach and its enemies*. New York: Basic Books.
Deffeyes, K.S. 2001. *Hubbert's peak*. Princeton, NJ: Princeton University Press.
The Economist. 2012. Call in the B team. October 6, p. 76.
Edwards, A.R. 2010. *Thriving beyond sustainability*. Gabriola Island, BC: New Society Publishers.
Forrestor, J.W. 1961. *Industrial dynamics*. Westford, MA: Pegasus Communications.

Heinberg, R. 2011. *The end of growth: Adapting to the new reality*. Gabriola Island, BC: New Society Publishers.

Heyd, T. 2007. *Encountering nature: Toward an environmental culture*. Aldershot, UK: Ashgate.

Jackson, T. 2011. *Prosperity without growth: Economics for a finite planet*. London: Earthscan.

Meadows, D.H., Meadows, D.L., Randers, J., and Behrens, W.W., III. 1972. *Limits to growth*. New York: Universe Books.

Meadows, D.H., Randers, J., and Meadows, D.L. 2004. *Limits to growth: The 30-year update*. White River Junction, VT: Chelsea Green Publishing Company.

10

Economic and Political Context

Introduction

Losing Control: The Emerging Threats to Western Prosperity (King 2010), *Endgame: The End of the Debt Supercycle and How It Changes Everything* (Maudlin and Tepper 2012), *The Reckoning: Debt, Democracy, and the Future of American Power* (Moran 2012), *Freefall: America, Free Markets, and the Sinking of the World Economy* (Stiglitz 2010)—the titles of these books eloquently capture the mood and reality of our times. There is a massive transfer in economic power taking place from rich, but heavily indebted, Western economies to emerging and developing economies.

> The OECD's projections for 2060 (at constant purchasing-power parities) show the impact of fast catch-up growth in underdeveloped countries with big populations. Economic power will tilt even more decisively away from the rich world than many realise. In 2011 the current membership of OECD made up 65% of global output, compared with a combined 24% for China and India. By 2060 the two Asian giants will have a 46% share of world GDP, the OECD members a shrunken 42%. India's economy will be a bit bigger than America's, China's a lot.
>
> Even so the Chinese and Indians will still be less well-off than Americans.... The same forecasts show GDP per person in China at 69% of that in America; in India it will be only 27%. And Americans will increase their lead over the citizens of some developed countries like France and Italy. (*The Economist* 2012a, p. 71)

At the same time the political landscape is changing from a uni-polar world dominated by the United States to a multi-polar world dominated by no one. The scale and rate of change is unprecedented in modern history, excepting perhaps immediately after World War II. Conventional economic and political theories, policies and practices no longer work. Governments will bumble and fumble their way through the next 30 years as they desperately grapple with a changing world over which, individually, they have little influence or control. We have entered a period of high *instability* and *uncertainty* (King 2010, Roubini and Mihm 2011).

Some idea of the magnitude of the problem can be gained by looking at what is happening to the United States, which has the world's most powerful military and is its largest economy. Fullbrook (2012) presents statistics showing the extent of the decline in the United States. He compares the United States to the other Organisation for Economic Cooperation and Development (OECD) countries, Australia, Austria, Belgium, Canada, Czech Republic, Denmark, Finland, France, Germany, Greece, Hungary, Iceland, Ireland, Italy, Japan, South Korea, Luxembourg, Mexico, Netherlands, New Zealand, Norway, Poland, Portugal, Slovak Republic, Spain, Sweden, Switzerland, Turkey, and United Kingdom. In his introduction Fullbrook says (p. 6), "Forty years of neoliberalism and plutonomy have taken their toll on the United States of America. Not so long ago the American way of life was the envy of the world and rightly so. Memories of America's once greatness at providing for the many a society for human fulfilment and happiness live on. But the times have changed. This book of 65 tables shows that in terms of the quality of life that it offers its citizens today, the USA is near the bottom of the third division of the thirty OECD nations". Overall it is ranked 29 out of the 30 countries, below Portugal and above Mexico. Table 10.1 summarises the U.S. rankings for the seven categories of indicators.

There are further problems: America's aging infrastructure and its large debt. The American Society of Civil Engineers 2009 report card on infrastructure shows a low overall grade of D (ASCE 2009). It estimates that $2.2 trillion is required to improve the infrastructure to grade A over 5 years. Needless to say, this level of investment has not been provided.

The United States needs to spend on infrastructure and improve the quality of life of its people at a time when it has incurred an astonishingly large debt that is still growing. Things cannot go on as before: it is endgame (Maudlin and Tepper 2012). America has to reduce its debt. At the same time, it is being challenged by the emerging economies, notably Brazil, Russia, India, China and South Korea (BRICK). American hegemony is on the wane, which will lead to an unravelling of military and political alliances (Moran 2012).

TABLE 10.1

U.S. Ranking against OECD Countries

Category	U.S. Rank (out of 30; lower is worse)
Health indicators	28
Family indicators	30
Education indicators	18
Income and leisure indicators	27
Freedom and democracy indicators	28
Public order and safety indicators	30
Generosity indicators	24

Source: Fullbrook, E., *Decline of the USA*, Bristol, UK: Real-World Economic Books, 2012.

Banking and Financial Markets

Modern economies depend on money. "Money is a medium of exchange, a unit of account, and a store of value" (Mayer 1997). We use money as medium of exchange, instead of direct barter, because it is easier. In principle, anyone can create money; the problem is getting self-created money accepted by others. Governments allow only their money to be used within their borders. This is called fiat money because it has been created by government command (fiat). Money is a unit of account in that it measures the comparative value of goods and services that people want to buy and sell. These two factors go hand in hand: in any economy the medium of exchange and the unit of account should be the same.

The final characteristic of money is that it is a store of value, allowing people to save for the future. Money works fine in this respect unless there is a period of very high inflation (money loses it value over time) or a period of uncertainty. If this occurs people will buy other things as a store of value, e.g., gold, cans of beans, commodities, etc. At the time of writing, financial investment experts are advising clients to buy gold and commodities to hedge against future changes in the value of money because of uncertainty in the financial markets and possible currency wars.

There is a bewildering variety of financial institutions that deal in money and related financial instruments. The most important of these are banks. The type of bank we are most familiar with (the commercial bank) has two main functions (Mayer 1997, Stiglitz 2010): (1) it provides an efficient payment system to transfer buyers' money to sellers, and (2) it collects money from depositors and packages these into loans for its borrowers. The money collected from depositors is on demand—it must be paid back any time the depositor wishes—and it must be repaid in full. Western banks make money by charging service fees for payments and interest on the money they loan.

A bank incurs liabilities when it obtains money to operate. The first way it obtains money is through issuing stock. The stock is bought by investors. "The bank doesn't 'owe' the shareholders that money in return, but it does owe them a share of the profits. That's why shares are considered liabilities: the shareholders have an equity claim on the bank" (Roubini and Mihm 2011). The second way a bank obtains money is by borrowing money through short-term deposits and loans.

The loans that banks make are considered assets "because they are investments that will, over the long run, give the bank a profit" (Roubini and Mihm 2011). The cash they hold is also an asset, as are the buildings and equipment they own. The difference between its assets and liabilities is the net worth of the bank (its capital or equity).

Western banks create money through a system known as fractional reserve banking, illustrated in Figure 10.1. Fractional reserve banking means that a bank must keep a certain proportion of deposits on reserve, but it can loan

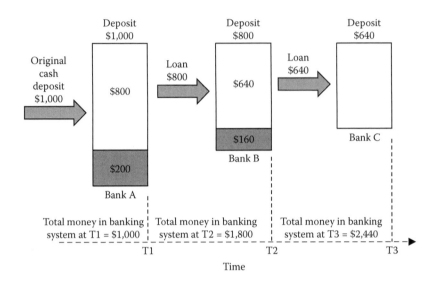

FIGURE 10.1
How fractional reserve banking creates virtual money.

out the rest. Assume you deposit $1,000 in Bank A at time T1. If the fractional reserve is 20%, then the bank sets aside $200, shown by the grey box, but is allowed to loan out the remainder ($800). The total amount of money in the system at time T1 is $1,000.

Bank A now loans out the remaining $800. Often the loan will be made to a depositor of the same bank, but for simplicity we will assume the borrower has an account at another bank, Bank B. It makes no difference, as the principle is the same. The $800 loan is accepted as a deposit at Bank B at time T2. The total amount of money deposited in the two banks is now $1,800, not $1,000. The additional $800 is virtual money created by the banking system.

Now Bank B sets aside 20% of the $800 ($160) and loans out the remaining amount of $640, creating more money at time T3—the total amount of money in the system is $2,440. If the process is repeated many times the total amount of money in the system will be $5,000. The original $1,000 has been multiplied five times.

In short, every time a bank issues a loan it creates virtual money. The amount of money created depends on the reserve ratio. The lower the ratio is, the greater the amount of virtual money created; e.g., 5% ratio multiplies the original deposit 20 times. It is not possible to separate the real money (notes and coins) from the virtual money because all deposits are aggregated and then packaged into loans. It is very important to note that banks accept deposits on short term but loan money long term. As can be seen from the illustration, banks borrow money from and loan money to each other. They do this not only within their own country but globally as well.

This type of banking system critically depends on trust. When you deposit money into your bank account the bank promises to pay you back in full. You trust the bank to honour its promise. But it is just that—a promise. If the bank makes mistakes, or the banker is a crook, you could lose some or all of your money in your bank accounts. Similarly, when the bank loans money to a borrower, the borrower promises to pay back the money. Banks need to cover the risk that borrowers cannot pay back their loans despite their promises to do so. Banks do this by asking borrowers to pledge assets to them on condition that they have the right to sell these if the loans are not paid back—assets assigned in this manner are called collateral. For example, if you have a bank loan (mortgage) on a property, then the property will be collateral for the loan. If the loan is for a company, then the company's physical assets (buildings, equipment, inventory, etc.) or its stock market value will be used as collateral. Collateral reduces loan risk provided the value that can be obtained when it has to be sold is the same as or more than the amount when it was pledged at the time the loan was issued. Depositors expect banks to be prudent in their lending practices and not to take undue risks. Good banking should be conservative and boring. Stiglitz (2010, p. 5) expresses the position succinctly:

> If a bank makes poor credit assessments, if it gambles recklessly, or if it puts too much money into risky ventures that default, it can no longer make good on its promises to return depositors' money. If a bank does its job well, it provides money to start new businesses and expand old businesses, the economy grows, jobs are created, and at the same time, it earns a high return—enough to pay back the depositors with interest and to generate competitive returns to those who have invested their money in the bank.

What happens if depositors do not believe a bank can or will pay back their money? Obviously they will rush to the bank as quickly as possible and demand their money—this is called a run on the bank. The bank cannot pay back all its depositors at the same time because only a portion of the deposits are held in reserve. It cannot easily get the money from its loans because the borrowers will not have the cash to pay them off. The bank could try to sell the borrowers' assets to recover the money, but this takes time. Furthermore, if a large number of assets are sold at the same time, their market value will drop and the full amount of the loans will not be recovered. In such situations banks can become insolvent.

In general, banks get into trouble in several ways. On the liability side of the balance sheet, they may not be able to borrow money or the interest rates may be too high. On the asset side of the balance sheet, their loan assets may drop in value. "When a bank's liabilities outstrip its assets, the bank's net worth sinks to zero. It is insolvent or bankrupt" (Roubini and Mihm 2011).

To help banks in trouble, some countries have established central banks. A central bank has three main functions: (1) to regulate and monitor banks,

(2) to assist banks that get into trouble, and (3) to stimulate or dampen the economy by adding or removing money from it. Central banks typically assist banks in trouble by loaning them money and guaranteeing deposits (up to a certain amount). As a depositor you will be reassured by this. There is a downside, though. It leads to irresponsible banking by encouraging bankers to take undue risks. Bankers will take risks because if they succeed, then they will gain a lot of money, and if they fail, the central bank will bail them out. There is no downside! This is a classic case of moral hazard and was one of the contributing factors to the 2008 financial crisis (Roubini and Mihm 2011, Stiglitz 2010).

Central banks create money out of nothing. "Let there be money" and there is. The money they create is normally issued to banks as loans; they are deposits from the banks' point of view. The banks, in turn, can lend out this money and the multiplier effect kicks in. For example, if a central bank lends $400 million to a bank with a fractional reserve ratio of 20%, then effectively the central bank is injecting $2 billion into the economy.

The banking system just described depends on the following two conditions to operate successfully:

1. Credit money: The system relies on people and companies borrowing money. This drives the whole system.
2. Economic expansion: Borrowed money must be paid back with interest, and this requires a growing economy over the long term.

The banking system fails when either condition is not satisfied.

Thus far, we have discussed commercial banks. However, there are other kinds of banks and financial institutions that are not as highly regulated as commercial banks, e.g., investment banks, hedge funds and insurance companies. These institutions add considerable risk and instability to financial markets.

An amusing article in *The Economist* (2012b, p. 61) titled "The Golden Rules of Banking" describes popular feelings about bankers:

> The crisis has taught people a lot about the banking industry and the thought processes of its leaders. These lessons can be distilled into four golden rules.
>
> 1. *The laws of supply and demand do not exist....* It is harder to be sure which trainee will be the next Nick Leeson and which the potential George Soros. This gives executives an excuse when things go wrong.
> 2. *Success is down to my genius; failure is caused by someone else....*
> 3. *What is lucky for an individual trader may be unlucky for the bank as a whole.* There is a survivorship bias in both fund management and trading. If your career starts with some bad losses, it will quickly come to an end. So, by definition, veteran traders will have had initial success. But that could be down to luck, not skill....

4. *Resigning can be a retirement plan.* When ordinary folk resign, they are lucky to get paid to the end of the month. But when bankers leave in awkward circumstances, they make out like a lottery winner.

The Capitalist Economy

System scientists distinguish between two different kinds of generic systems: concrete and abstract. Concrete systems have substantive components: people, machines, factories, etc. Social systems are concrete systems whose components are people or other social systems.

Technological systems are concrete systems whose components are tools, machines and equipment. Some researchers include people in the technological system, but strictly speaking this is not correct. Consider, for example, a person loading and unloading materials and parts from a machine. From a technological perspective, what is important is the movement of materials and parts and not the object that performs the movement. The movement could be performed by a robot and achieve the same results. The robot and the person are interchangeable in this respect. In other words, a person is not required to perform the movement *as a person*.

A socio-technical system is a system that contains a social system and a technological system (Chapter 14), e.g., a SMMC. Companies and other organisations are either social systems, socio-technical systems or some combination of these.

The components of abstract systems are concepts and ideas: mathematical concepts, philosophical concepts, etc. Abstract systems are belief systems. Contrary to popular belief, economic and political systems are abstract systems, not concrete (real systems). A political system is created by viewing the interactions between people, social systems or socio-technical systems in terms of control—how they should be governed.

An economic system is created by viewing the interactions between people, social systems and socio-technical systems as transactions. It is interesting to note that originally economics was considered a branch of moral philosophy, confirming its role as an abstract system. Some economists and researchers, including the authors, argue that this is economics' proper role. Basically economics is concerned with the efficient allocation of scarce resources. It states what one ought to do to use resources efficiently. Economics is also concerned with the just and equitable distribution of these scarce resources, thus bringing in politics.

Capitalism is an economic system that is based on private ownership of social and socio-technical systems to produce goods and services for profit. Each privately owned system generates and distributes wealth at the same time: these are its two main functions. Economists argue that the capitalistic market is the most efficient allocator of resources. Their mathematical models determine the most efficient allocations under different assumptions and conditions. They also believe the market is efficient at equitable distribution despite considerable evidence to the contrary, which is why governments often intervene to redistribute market-generated wealth. Modern economists seem to have overlooked that economics is rooted in politics because scarcity involves choice and choice involves politics (King 2010). The *political economy* is important, not the economy: "as the emerging nations stake their claim on the world's scarce resources, the West's influence on global economics affairs is on the wane. How will scarce resources be distributed? Will the market be left in charge or will nation states increasingly call the shots? … The interaction between the Western world and the emerging nations brings the politics back into economics (and, for that matter, takes the mathematics out)" (King 2010, p. 10).

Equilibrium economists argue that the capitalist market is self-equilibrating despite the many market failures that have created widespread unemployment and hardship. Equilibrium economists blame the market collapses on extraneous factors and not on the underlying structure of capitalism. When a crisis occurs they say *this time is different*. But they are wrong!

Crisis economists argue that the market structure is inherently unstable; therefore boom and bust cycles are inevitable. Two leading economists, Reinhart and Rogoff, have demonstrated very clearly that the underlying structure of capital markets is unstable. Their critical analysis covers eight centuries of financial crises of all types: inflation, currency crash, currency debasement, banking, external debt and internal (domestic) debt. Table 10.2 shows the number of banking and external debt default/rescheduling crises that have occurred. On average there have been three banking crises every 2 years since 1945. Stable markets indeed!

For our purposes we will focus on two types of crises: banking and external debt. Banking crises are important because they occur in all economies (developed and emerging) and tend to be global, either regionally or widespread. Reinhart and Rogoff refer to them as an "equal-opportunity menace". Banking crises spread globally because banks loan money to each

TABLE 10.2

Number of Banking and External Debt Default/Rescheduling Crises

Number of External Debt Default/Rescheduling Crises (from independence to 2008)	Banking Crises	
	Number since Independence or 1800	Number since Independence or 1945
240	268	106

Source: Reinhart, C.M., and Rogoff, K.S., *This Time Is Different: Eight Centuries of Financial Folly*, Princeton, NJ: Princeton University Press, 2009, Tables 6.5, 6.6, 10.3 and 10.4.

other, not only within their own country but across borders. "Periods of high international capital mobility have repeatedly produced international banking crises, not only famously, as they did in the 1930s, but historically" (Reinhart and Rogoff 2009, p. 155).

External debt crises are important because they frequently occur in emerging economies as they develop. An external debt crisis occurs when a country defaults on its payments to its external creditors. The largest external debt crisis occurred in 2001 when Argentina defaulted on US$95 billion. This type of default occurs because there is no international legal framework for creditors to enforce sovereign debt contracts.

Reinhard and Rogoff (2009, pp. 270–273) propose a prototypical pattern for the development of a financial crisis:

- Stage 1: A period of financial liberalisation. Money markets become less regulated leading to all kinds of financial innovation.
- Stage 2: Beginning of a banking crisis.
- Stage 3: Currency crash.
- Stage 4: Inflation increases.
- Stage 5a: Peak of banking crisis if country does not default on debt.
- Stage 5b: Default on debt (external or domestic).
- Stage 6: If default occurs (5b), then inflation crisis worsens and this is the peak of the banking crisis.

From stage 2 onward there are stock market and property market crashes accompanied by a slowdown in the economy. From stage 3 onward capital controls are likely to be used as countries try to shield their economies from rapid inflows and outflows of capital.

We will use the 2008 financial crisis to illustrate how stage 1 builds up and leads to stage 2. Stage 1 starts with a relaxing of financial regulations allowing innovative financial products to be created. What was relaxed in this case? The way home loans were created and managed in America. In the old model, banks created loans and *held* onto them. With financial innovation banks created loans and then *distributed* them. Mortgages were pooled and sold as bonds or other kinds of financial instruments: "illiquid assets like mortgages could now be pooled and transformed into liquid assets that were tradable on the open market" (Roubini and Mihm 2011). The name given to this process is securitisation. Mortgages were only the start. Soon all kinds of loans were being securitised—consumer loans, e.g., credit card loans—and "by the time the crisis hit, securitisation had been applied to airplane leases, revenues from forests and mines, delinquent tax liens, radio tower revenues, boat loans, state and local government revenues, and even the royalties of rock banks" (Roubini and Mihm 2011, p. 65). The innovative packaging of these securities meant banks could not properly assess the risks when they bought them.

Financial innovation encouraged banks to take risks. The pressure to issue loans increased, and so consumers were encouraged to borrow more and more. As they borrowed more, housing and other asset prices increased due to the increased money in circulation. This allowed consumers to borrow even more against their assets. Interest rates were deliberately kept low by the Federal Reserve, and this encouraged even more borrowing.

Money was needed to drive all this. Where did it come from? It came from overseas. Money poured into the United States despite the low interest rates. The reason for this was most that most of the money was provided by sovereign wealth funds whose managers were looking for a safe haven for liquid assets. They were not worried about the low returns. The consumer boom was fed by America going into debt: public and private debt. In system terms this is a deviation-amplifying process. It will continue until some critical limit or event occurs and then the system crashes:

> When the bubble popped, the effects were amplified because banks had created complex products resting on top of the mortgages. Worse still, they had engaged in multibillion-dollar bets with each other and with others around the world. This complexity, combined with the rapidity with which the situation was deteriorating and the banks high leverage (they, like households, had financed their investments by heavy borrowing), meant that banks didn't know whether what they owed to their depositors and bondholders exceeded the value of their assets. And they realized accordingly that they couldn't know the position of any other bank. The trust and confidence that underlies the banking system evaporated. Banks refused to lend to each other—or demanded high interest rates to compensate for bearing the risk. Global credit markets began to melt down.
>
> At that point, America and the world were faced with both a financial crisis and an economic crisis. (Stiglitz 2010, p. 3)

The crash is also deviation amplifying but in the opposite direction. The recovery from severe financial crises is long and painful (all numbers below are averages based on empirical data):

- Asset markets collapse. Real housing prices decline 35% for 6 years. Equity prices decline 56% for 3.5 years (Reinhart and Rogoff 2009).
- Output and employment decline. Unemployment rises 7% and lasts for the 4-year duration of the down phase of the cycle. Output drops 9% (from peak to trough) for 2 years (Reinhart and Rogoff 2009).
- Government debt increases dramatically: "on average 86% (in real terms, relative to precrisis debt) in the major post–World War II episodes" (Reinhart and Rogoff 2009, p. 224). The rise in debt is mainly due to massive reductions in tax revenue. The bailout costs for the banking system are small in comparison to this (Reinhart and Rogoff 2009).

- De-leveraging (debt reduction) lasts 6–7 years (Maudlin and Tepper 2012).

Financial crises might be triggered by specific events, but they occur because financial and economic systems are inherently unstable. Instability can be mitigated by adopting the following practices:

- Allow local exchange trading systems (LETSs) (Heinberg 2011). LETSs support and promote local communities and attenuate the adverse effects of fiat currencies. LETSs are based on local currencies that have the following features:
  - The currency (money) can be created by anyone.
  - Each currency will serve a different function in society.
  - No interest is charged on the money.
  - The money is simply a mutual credit agreement between a buyer and a seller. It may or may not have physical representation.
  - The money should be backed with something tangible so that it can serve as a unit of account.

  A LETS is a system "in which each transaction is recorded as a corresponding credit and debit in the two participants' accounts. The quantity of currency issued is always and automatically sufficient and does not depend on a bank or Government for issuance" (Heinberg 2001, p. 243).
- Establish a banking system with 100% reserve ratio and not allow banks to charge interest. Under these conditions banks would return to their original role of loan and hold. Not charging interest preserves the stored value of money over time and reduces competitive income behaviour.

Income Inequality

The standard measures economists use to measure inequality clearly indicate that global inequality (between countries) is decreasing; that is good news. The bad news is the inequality within nations is increasing in most countries; the rich get richer and the poor get poorer. There has been a dramatic increase in inequality within nations over the last 30 years, whereas previously it had been decreasing:

> Including capital gains, the share of national income going to the richest 1% of Americans has doubled since 1980, from 10% to 20%, roughly where it was a century ago. Even more striking, the share going to

the top 0.01%—some 16,000 families with an average income of $24m—has quadrupled, from just over 1% to almost 5%. That is a bigger slice of the national pie than the top 0.01% received 100 years ago.

This is an extraordinary development, and it is not confined to America. Many countries, including Britain, Canada, China, India and even egalitarian Sweden, have seen a rise in the share of national income taken by the top 1%. The numbers of ultra-wealthy have soared around the globe. (*The Economist* 2012c, p. 3)

The world's two largest economies, America and China, one developed and one developing, have approximately the same degree of high inequality (roughly 0.4 as measured by the Gini coefficient).

Traditionally economists have argued that income inequality is essential for efficient markets. This belief has been challenged by recent studies that indicate inequality leads to inefficient markets due to cronyism, rent seeking (someone taking more money than he or she is worth), and lack of upward mobility, which in turn inhibits innovation.

Governments typically intervene to redistribute wealth. Two basic strategies are followed: equality of opportunity and equality of outcome. "Europeans tend to be more egalitarian, believing that in a fair society there should be no big income gaps. Americans and Chinese put more emphasis on equality of opportunity. Provided people can move up the social ladder, they believe a society with wide income gaps can still be fair" (*The Economist* 2012c, p. 6). Recent studies indicate rising inequality produces less upward mobility; therefore governments need to consider how to ensure equality of outcome, but in a way that does not impair market efficiency.

Global Politics

A sovereign state is a society with a central government that has control over a specific geographic area. Sovereign states may contain people of one ethnic group, culture or religion (nation-state), or may contain several (multi-nation state). Singapore is an example of a multi-nation state. It has three major ethnic groups: Chinese, Malay and Indian. There has been a rapid rise in the number of sovereign states in the last hundred years. In the 19th century the world consisted of a few empires and sovereign states. Today there are over 200 sovereign states. Politically, the world has been disintegrating into smaller and smaller political entities. Correspondingly, the number of national banking systems and markets has increased.

Global political instability and uncertainty is likely to occur over the next 30 years because of conflict over scarce resources, financial market uncertainty and increased demand for global democracy. Factors contributing to instability and uncertainty include the following:

- National security. Countries will attempt to gain control of or access to scarce resources to protect their people, and they are likely to erect trade and immigration barriers to protect their economies. This will lead to increased trade and resource conflicts.

- Currency wars. The rich nations will most probably de-leverage their debt by printing money and hence devaluing their currencies, as is happening now in the United States and Europe. Accordingly, currencies of other countries will rise in value. This has two important effects. If a country is mainly an exporting country, then its exports become more expensive. Its trading position is threatened. If a country has high interest rates or its currency is seen as safe, there will be capital inflows ("hot" money) from the devaluing countries. These flows can be large compared to the size of the economy. The only sensible recourse for countries affected in this way is to impose capital controls.

- Global democratic deficit (Stiglitz 2007). Sovereign states are demanding a say in world affairs. The current world institutions and forums are heavily biased toward a few Western nations. Developing countries are asking for representation in these institutions. Some changes have already started; e.g., the G7 has expanded to G20. There are real challenges ahead:

  - Obtaining agreement from over 200 states on important issues
  - Obtaining agreement from Western and politically powerful states to concede power gracefully
  - Obtaining agreement from emerging powers to be patient in gaining power and to use peaceful means to resolve international disputes
  - Obtaining agreement from the United States to use its military power to stabilise the transition so that all sovereign states benefit, not just the United States alone, and from these states to support America in this role

- Enduring government legitimacy. Many governments are struggling to maintain long-term legitimacy.

  - People in non-democratic countries are increasingly rising up against their political masters and demanding democracy. This will lead to demonstrations, riots and perhaps civil war, as events in the Middle East have shown.
  - Fledgling democracies also face a legitimacy crisis. Many are multi-ethnic, multi-cultural or multi-religious nations. Conflicts between these groups can lead to political power

struggles. The government may change quickly and erratically. Furthermore, the power struggles can lead to demonstrations, strikes and riots, as has happened in Thailand.

- Mature democracies are not immune either. We have noted the increased inequality in some rich countries. The rich typically use their wealth to manipulate the political system to their advantage. At some point the disadvantaged majority will reach a breaking point and demand changes. Again there are likely to be demonstrations, strikes and riots.

How will nations respond to simultaneous economic, political, military and environmental turbulence? The most likely scenario is they will select one or more of the maladaptive social responses, particularly segmentation, leading to increased local and regional conflicts. The transition in economic, political and military power will not be easy. There will be winners and losers (King 2010, Stiglitz 2007):

> For much of the world, globalization as it has been managed seems like a pact with the devil. A few people in the country become wealthier; GDP statistics, for what they are worth, look better, but ways of life and basic values are threatened. For some parts of the world gains are even more tenuous, the costs more palpable. Closer integration into the global economy has brought greater volatility and insecurity, and more inequality. It has even threatened fundamental values.
>
> This is not how it has to be. We can make globalization work, not just for the rich and powerful but for all people, including those in the poorer countries. (Stiglitz 2007, p. 292)

Guidelines for Regenerative Development

Past and recent actions by governments in both developed and developing nations do not encourage us to believe they will take any significant action to move their economies toward long-term sustainability. The following guidelines are intended to help SMMCs navigate the future political and economic uncertainties:

- Go global. Globalisation is a hedge against uncertainty. The current banking and debt crisis affects countries to different degrees. If markets are down in one country, they may well be up in another. Where possible, manufacture and obtain supplies from the same region where the market is. This is a hedge against interruptions to global supply chains and currency exchange fluctuations, and promotes the development of local communities.

- Know what is happening. Rapid and frequent financial, economic and political changes should be expected. Rapid response information systems need to be set up to detect and report these events. CEOs need to ensure they understand the longer-term implications of the changes. We recommend they read *The Economist* and the works of crisis economists and political commentators.

- Do not trust banks. The banking crisis is not over yet. When banks are in trouble they will stop short-term credit, stop loans and require companies to make up the shortfall when loan covenant limits are breached. For all these reasons, banks should not be trusted. Keep sufficient cash reserves to survive a credit crunch. Fund development projects from profit and cash reserves. If you have loans from a bank, then prepare contingency plans to repay some or all the loan at short notice. Under no circumstances offer equity to any financial institution in order to obtain a loan.

- Incur capital debt in the countries where the markets are and in proportion to the market value as far as possible. This reduces currency exchange risk.

- Implement and promote flexible work practices.

- Implement employment conditions that limit income inequality. Promote equality of outcomes in preference to equality of opportunity.

Exercises and Problems

1. Briefly describe how fractional reserve banking works.
2. What is the fractional reserve ratio for banks in the country in which you reside?
3. Briefly describe Richard Branson's Carbon War Room. (It is not described in this book. You will have to do your own research.)

References

ASCE. 2009. 2009 grades. www.infrastructurereportcard.org (accessed October 14, 2012).

The Economist. 2012a. November 10, p. 71.

The Economist. 2012b. The golden rules of banking. July 14, p. 61.

The Economist. 2012c. For richer, for poorer. Special Report, World Economy. October 13, p. 3.

Fullbrook, E. 2012. *Decline of the USA*. Bristol, UK: Real-World Economic Books.

Heinberg, R. 2011. *The end of growth: Adapting to the new reality*. Gabriola Island, BC: New Society Publishers.

King, S. 2010. *Losing control: The emerging threats to western prosperity*. New Haven, CT: Yale University Press.

Maudlin, J., and Tepper, J. 2012. *Endgame: The end of the debt supercycle and how it changes everything*. Hoboken, NJ: John Wiley.

Mayer, M. 1997. *The bankers*. New York: Truman Talley Books/Dutton.

Moran, M. 2012. *The reckoning: Debt, democracy, and the future of American power*. New York: Palgrave Macmillan.

Reinhart, C.M., and Rogoff, K.S. 2009. *This time is different: Eight centuries of financial folly*. Princeton, NJ: Princeton University Press.

Roubini, N., and Mihm, S. 2011. *Crisis economics: A crash course in the future of finance*. London: Penguin.

Stiglitz, J.E. 2007. *Making globalization work*. New York: Norton and Company.

Stiglitz, J.E. 2010. *Freefall: America, free markets, and the sinking of the world economy*. New York: Norton and Company.

Postscript to Section II

The Uniqueness of Moral, Aesthetic and Ecological Experiences

The significant others of morality are not interchangeable. Each moral encounter with a significant other is a unique experience. Aesthetic objects are not interchangeable. Each aesthetic experience is a unique event. Ecologies and community ecological systems (CESs) are not interchangeable. Their histories, locations and interrelationships are unique. Life in them is unique.

Herein lies the major difference between these experiences and those explained by modern science. Science aims at generality, not uniqueness. Economics aims at generality, not uniqueness. Moral, aesthetic and ecological experiences are outside the scientific domain by definition—there cannot be a science of the unique.

Science and economics dominate our current thinking and worldview to the detriment of design and the arts. There is an urgent need to balance this dominance with unique views and appreciations of the world. By so doing, people will enrich their lives and live in harmony with nature. This, coupled with the fact that the world economy cannot keep growing forever, means that our current economic and political systems will have to change. How can we change them?

The most effective approach for changing systems is through an idealised design (Chapter 16). An idealised design is a design of a system that stakeholders would like to have. It must be technologically feasible and capable of surviving if implemented. The remainder of the postscript presents a list of features we believe an idealised design of a regenerative, global political economy should have. The design is intended to encourage pursuit of all four ideals.

An Idealised Design for Regenerative Development

The aims of the design are to:

- Provide a vision of a stable political economy, which can serve as a guide for planning
- Encourage designers and SMMCs to design and experiment with different forms of economic and political organisation at work, locally in their communities, and regionally.

The design assumptions and features have been adapted from the works of ecological economists and other writers. All but one of the design features have been implemented in some form or other, so the design is feasible. The exception is democratic governance globally.

The assumptions underlying our proposal are:

1. Planet earth is finite (Chapter 9).
2. We are citizens of the world. World citizenship should take priority over national citizenship.
3. The earth's resources belong to everyone. Everyone, rich and poor, is entitled to their fair share of the resources. Furthermore, we are temporary custodians of these resources. We have a duty to ensure the world is left in a better state for future generations; they are entitled to their fair share.
4. There is only one economy: the world economy.
5. There is a steady-state world population at a sustainable level (Chapter 9).
6. There is a steady-state (zero-growth) world economy. The assumption that national economies and the world economy can keep growing forever in a world of finite resources is completely and utterly unrealistic. There are only two ways economies can grow: by an increase in population or an increase in productivity. Chapter 9 showed that in the long run the earth has finite carrying capacity, and at some stage there will be an end to global population growth. Chapter 8 showed that there are limits to increasing productivity. We have already reached productivity limits in some industries, and this trend will be repeated in other industries.

Money and Banking

- Allow and encourage local exchange trading systems.
- Banks will have a reserve ratio of 100% and not be permitted to charge interest.

The Global Economy

- Cooperation before competition. Unfettered competition is inherently destructive. Cooperation improves learning, and makes people happier, more secure and more productive (Kohn 1986). Many studies have shown the superior performance of cooperatives compared to companies organised competitively. Note also that cooperation is essential to stabilise turbulent social fields (Chapter 7). Therefore the economy will be based on cooperatives.
- Cooperatives can be for profit and not for profit. They will be designed according to the following principles (Heinberg 2011, p. 254):
  - Voluntary and open membership
  - Democratic member control
  - Member economic participation
  - Autonomy and independence
  - Education, training and information (of its members)
  - Cooperation among cooperatives
  - Care for the community
- Competition between companies or cooperatives will be embedded within a cooperative legal framework designed to encourage moral competitive behaviour.
- The world economy will be based on CESs.
- CESs will be based on small and medium cooperatives. Research has shown that small and medium companies generate more jobs than large ones and distribute income more equitably. Cooperatives can be set up as companies, though they need not be.
- The economy will be designed as an open system with real-time control. Waiting months for economic data is ridiculous given modern technology. The economy needs to be modelled in a manner similar to processes in companies. We are aware of only one example where such modelling and real-time control has been achieved; it was in Chile during President Allende's tragically short term (Beer 1975, 1981).
- Specify company law to ensure companies pay the true costs of resource depletion and environmental pollution (Heinberg 2011). Currently companies can avoid these costs through their limited

liability. They should be responsible for the full life cycle costs of their products.

- Tax environmental "bads," such as resource depletion and pollution, in preference to company outputs (Heinberg 2011; Jackson 2011).
- A system of tariffs "that will allow countries that implement sustainable policies to remain competitive in the global marketplace with countries that don't" (Heinberg 2011, p. 252).
- Ecological investment. Invest in projects that will improve the ecology (Jackson 2011).
- Economic measures based on ideals, e.g., gross national happiness.
- A zero-growth world economy based on developing human resources through lifelong learning.
- Flexible workdays.
- No forced retirement age. People will be allowed to work as long as they want to.

Inequality

- Reduce inequality between nations by:
  - Restricting or banning immigration controls
  - Restricting or banning tariffs and other trade barriers, except for those relating to regenerative development policies and practices
- Reduce inequality within nations by adopting "true progressivism" (adapted from *The Economist* 2012).
  - Restrict cronyism and promote competition (to prevent rent seeking): legislate and take action against political corruption, and legislate against large monopolies and oligopolies.
  - Social spending should be targeted and progressive. Use conditional cash transfers to focus spending on social objectives, such as universal education for the young or upgrading skills to meet changing economic conditions. Conditional cash transfers are direct payments to the needy provided they meet certain conditions, such as sending their children to school. In rich countries, spending needs to be directed away "from older and richer people to younger and poorer ones" (*The Economist* 2012, p. 27).
  - Set limits on income inequality. Promote equality of outcomes in preference to equality of opportunity.

The Global Political System (Heinberg 2011, Stiglitz 2007)

- Democratic governance within nations and globally.
- Enforceable international legal framework, especially for the two most precious resources: people and ecosystems.

- Common security clubs. Social action at the local level to promote and encourage people to adopt and maintain regenerative development practices.

SMMCs

SMMCs will be designed to actively encourage and support their stakeholders to seek ideals, as follows. SMMCs will:

- Develop products and environments that are aesthetically positive and refrain from developing those that are aesthetically negative.
- Relate to their stakeholders in ways that improve their aesthetic capability.
- Relate to their stakeholders in ways that improve their moral behaviour and ethics.
- Promote and implement democratic decision making internally and externally.
- Support CESs and LETSs.
- Develop the technical capabilities of their stakeholders through lifelong learning.
- Promote and support indigenous knowledge and understanding as well as that generated through modern science methodologies.
- Nurture the environment through ecological investment.

References

Beer, S. 1975. *Platform for change*. New York: John Wiley.
Beer, S. 1981. *Brain of the firm*. 2nd ed. New York: John Wiley.
The Economist. 2012. The golden rules of banking. July 14, p. 61.
Heinberg, R. 2011. *The end of growth: Adapting to the new reality*. Gabriola Island, BC: New Society Publishers.
Jackson, T. 2011. *Prosperity without growth: Economics for a finite planet*. London: Earthscan.
Kohn, A. 1986. *No contest: The case against competition*. Boston: Houghton Mifflin Company.
Stiglitz, J.E. 2007. *Making globalization work*. New York: Norton and Company.

Section III

Synthesis—Technologies

11

Life Cycle Analysis

Introduction

A life cycle is a model of a repeatable process, which is the complete life cycle of a system. For a biological system the process starts with conception and ends when the system dies. For a designed system the process starts from the time it is conceived and ends when it is retired. The process is divided into phases or stages; for example, a software development life cycle might be divided into specification, design, coding and assurance. Each phase requires different methods and skills. The phases may overlap and iterate, but usually start in the sequence shown in the life cycle.

Life cycles are used to reduce the time and cost in setting up new projects and processes; to improve learning, quality and risk management; and to facilitate correct allocation of resources to the different phases of system development. Learning is improved through repetition and quality is improved through learning and compliance to procedures. Risk management is improved because the model ensures critical activities are identified and controlled and that feedback from prior cycles is used to improve future activities. The main disadvantage of using a life cycle is it may inhibit process innovation.

The concept map for this chapter is depicted in Figure 11.1. The chapter presents a general system life cycle model. It describes some of the main cycles used in industry to develop products and manufacturing systems and shows how intersecting life cycles can harmonise two or more development processes. The cycles do not include technology development, as it is more effective for technologies to be developed separately from the design-manufacture life cycle, as discussed in Chapter 3.

A generic life cycle that can be applied to all systems is shown in Figure 11.2. The cycle starts with the design phase. It is here the system is first conceived. During this phase the client requirements are generated and the system is conceived, designed and critiqued. Planning and organisation of the system development project is included in this phase.

The next phase is manufacturing for movable systems or construction for fixed systems. Critique is undertaken concurrently with the design and

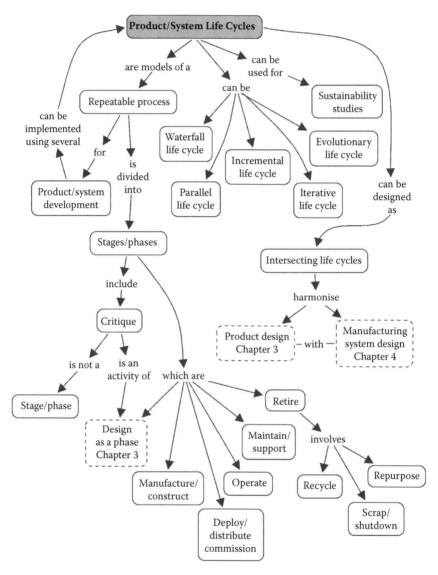

FIGURE 11.1
Concept map for Chapter 11.

manufacturing phases to ensure the system meets client needs and design requirements. This is followed by deployment and distribution of movable systems, and commissioning of fixed and movable systems. Critique is undertaken concurrently with this phase to ensure the system meets client needs and design requirements.

Next is use of or operation of the system by the end user. Maintenance and support activities may run in parallel with this phase. Critique may be

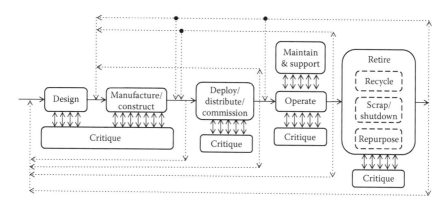

FIGURE 11.2
Generic life cycle of a system.

carried out during operations to prove that the system meets operational requirements and client needs. Finally, the system is retired. It is either recycled, scrapped or shut down and left, or repurposed. Critique may be carried out to prove that the system is phased out according to the disposal requirements.

The life cycle shows the feedback loops from later phases back to earlier phases (dashed lines). In later diagrams these feedback loops will not be shown for clarity, but you must remember that they should be there. Feedback improves a design of a new system based on manufacturing and operational experience of earlier designs of different systems. If prototyping is used, then feedback can occur within a design cycle to improve the design concept.

Brooks (2010) criticises this form of representation because it does not show how manufacturing and operational experience are used to design a new system. People from manufacturing and operations should be members of the design team and actively involved in designing the system. That is, they are also designers of the system. The feedback loops do *not* represent this kind of involvement and are not intended to.

The boundaries between the phases are often defined by checkpoints called *reviews*, some of which are *control gates* or *tollgates*. The purpose of a review is to assess progress before moving on to the next phase of development, though some reviews may be carried out during a phase. Table 11.1 lists the system development phases and reviews for two different life cycles. One is NASA's life cycle for one-off, complex space systems, and the other is General Electric's cycle for the design, manufacture and marketing of large batch and mass-manufactured products. The GE cycle is unusual as the phases are named after the tollgates, so there is no distinction between these. Reviews are not shown in Figure 11.2 for clarity, but they are necessary.

TABLE 11.1

Phases and Reviews for Two Different Life Cycles

NASA's Technical Reviews		General Electric Tollgates
Life Cycle Phase	**Review**	**Life Cycle Phase/Review**
Mission feasibility	Mission concept review	Customer/consumer needs
Mission definition	Mission definition review	Concept review
System definition	System requirements review	Preliminary design review
System design review	Feasibility review	Final design review
Preliminary design	Preliminary design review	Critical producibility review
Final design	Critical design review	Market/field test review
Fabrication and integration	Systems acceptance review	Manufacturing feasibility review
Preparation for deployment	Flight readiness review	Market readiness review
Deployment and operational verification	Operational readiness review	Market introduction and follow-up
Mission operations	Decommissioning review	
Disposal		

Source: Shisko, R., *NASA Systems Engineering Handbook,* Washington, DC: NASA, 1995; Clark, K.B., and Wheelwright, S.C., *Managing New Product and Process Development*, New York: Free Press, 1993.

Development Life Cycles for Industrial and Consumer Products

The simplest life cycle for designing, making and distributing products is the waterfall cycle, illustrated in Figure 11.3. The figure shows the design phase and not the other phases in the cycle. The other phases will be performed serially. (manufacture-deploy-operate-retire).

Most texts in systems engineering and design divide the design phase into several separate phases, e.g., conceptual design, embodiment design, detailed design, integration, validation and verification. Specification is assumed to be a separate phase from design. On the other hand, design experts argue that specification should not be separated from design. Our aim in this chapter is to show how the different views of the design process can be reconciled through different life cycle models. We will use our model of the design phase (Chapter 3) to illustrate the different design life cycle models. The design phase has four main activities: ideation, development, production and critique.

In the waterfall life cycle, the activities are executed in a serial manner. Each activity is performed once and completed before moving on to the next. Critique is performed at the end. The waterfall cycle is suitable for small development projects, but not for large ones.

The development time can be significantly reduced by overlapping the activities (Figure 11.4). This is called a parallel life cycle. Each activity can be divided into a number of smaller sub-activities, allowing a following activity

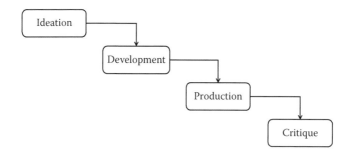

FIGURE 11.3
Waterfall life cycle.

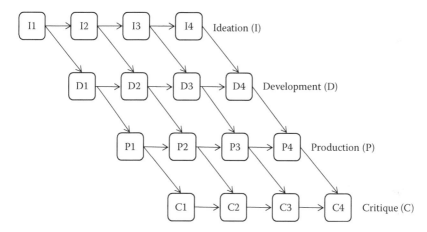

FIGURE 11.4
Parallel life cycle.

to begin before the previous one is completed. So instead of one large water-fall, there are several smaller waterfalls cascading together. The four main activities of Figure 11.3 are divided into four sub-activities numbered 1–4. The cycle starts with I1. Once this is completed I2 and D1 can begin and run in parallel. When I2 and D1 are completed I3, D2 and C1 can begin and run in parallel, and so on. The parallel life cycle is suitable for all products (hardware and software).

A parallel life cycle is possible only if two minimum conditions are met. First, the product architecture must be decomposable. That is, the product has to be designed such that it can be divided into sub-systems that can be manufactured independently. As with all decomposable systems, the interfaces have to be carefully designed to ensure the sub-systems can be integrated with minimum impact on product performance. Second, the over-all architecture has to be confirmed early in the design phase because the cost of a design change increases the later changes are made.

Three very effective life cycles are used in industry to develop software: incremental, iterative and evolutionary. In an incremental life cycle the finished product is delivered in several versions, with each version containing more functionality than the previous one (Figure 11.5). Each product version undergoes the complete life cycle once, but the activities are repeated between versions. It can be used where a product has clearly defined functionality that can be delivered separately in modules to the customer. The complete product has functionalities A, B and C. It is delivered in three versions. The first version has functionality A. The second has functionalities A and B. The third version has functionalities A, B and C.

It is possible also to use an incremental cycle for some hardware products. Usually this is achieved through product family or modular architectures when incrementing for customers. It is possible to increment development internally using intermediate forms of the product or system, as discussed later.

An iterative life cycle model is one where the whole system will repeat some of the activities a number of times. In essence, it is a prototyping cycle. It is commonly used for software development, but can also be used for hardware development when the failure costs of a system are high, as in a satellite; the cost of prototyping is low; or when the cost of prototyping is high and the number of products is very large, so the burden per customer is acceptable. The cycle is illustrated in Figure 11.6, which depicts three iterations. Normally ideation

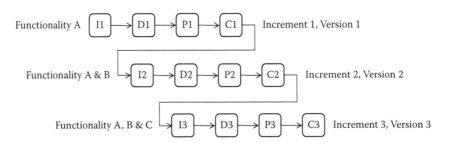

FIGURE 11.5
Incremental life cycle.

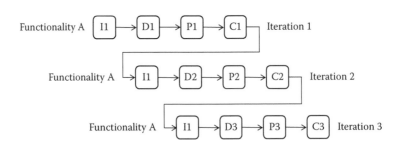

FIGURE 11.6
Iterative life cycle.

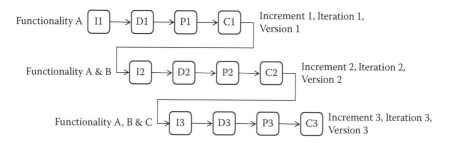

FIGURE 11.7
Evolutionary life cycle.

(specification and concept architecture) does not iterate. This can be seen in Figure 11.6; I1 is common to all iterations. Another feature of iteration is that the earlier prototypes may be discarded and new ones created; thus the embodiment architecture may iterate. The iterative life cycle is also referred to as a spiral life cycle. The incremental and iterative life cycles are often combined.

Recent advances in software development have led to a more radical approach: the evolutionary life cycle (Figure 11.7), commonly referred to as agile software development (Shore and Warden 2007). It is an incremental and iterative approach in which *all* activities of a software life cycle are iterated. The work is divided into small increments. Each increment is defined for the next cycle at the end of an iteration and delivers software the customer can use. The iterations are short, typically 1–4 weeks.

The cycle has some other interesting features. The requirements can change at any time during the cycle, developed software is frequently thrown away and replaced by new software, and the teams are self-organising, with programmers working in pairs. This kind of software development is best suited to situations where the requirements are ill defined or the customer has difficulty articulating needs. Currently it is used solely for software development; however, Alexander (2002) suggests that it might be possible to construct buildings using this approach. The authors agree and believe that an evolutionary approach may be suitable for integrating designed systems with natural ecologies.

Development Life Cycle for a Manufacturing System

Manufacturing systems are usually developed using a waterfall or parallel life cycle. The major phases are shown in Figure 11.8. Manufacturing systems consist of networks of suppliers connected together to deliver products to customers, and other essential stakeholders, such as waste removal companies, energy suppliers, water suppliers, etc. Therefore they are designed

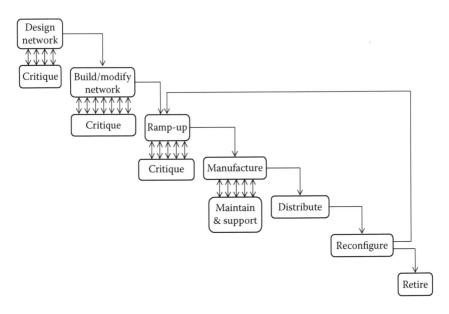

FIGURE 11.8
Manufacturing system life cycle.

as networks. The manufacturing phase refers to the products made by the manufacturing system and not the manufacturing system itself.

The cycle has two new phases not included in previous diagrams: ramp-up and reconfiguration. Ramp-up is a phase where manufacturing of a product begins, starting from a low level of capacity (say 20%) and gradually building up to full capacity (100%). The purpose of ramp-up is to mitigate risk. Employees have time to identify manufacturing problems and constraints and to develop and implement solutions. Once full capacity is achieved management can be confident that the manufacturing process will be stable and operating at the expected performance levels.

The other phase is reconfiguration. There are two basic approaches to flexible manufacturing. The first is to pre-design a factory so it can accept families of products. Later product variations are incorporated without changing the layout. The best example of this approach is group technology. The second approach is to design a factory so it can be redesigned and reconfigured quickly and economically for new products, hence the extra phase.

Choosing a Life Cycle

The life cycle comparison chart (Table 11.2) can be used to select the life cycle that would be appropriate for a particular system development

TABLE 11.2
Life Cycle Comparison Chart

	Waterfall	Parallel	Incremental	Iterative	Incremental and Iterative	Evolutionary
Applicability	All systems	All systems	All software; some physical products and manufacturing systems	All software; some physical products and manufacturing systems	All software; some physical products and manufacturing systems	Some software
Cycle duration	Less than 6 months	Greater than 6 months	Greater than 6 months	3–4 weeks	Greater than 6 months	3–4 weeks
Customer needs and requirements	All known, certain and fixed	All known, certain and fixed	Some known, certain and fixed; some not known or not certain and defined later	Most or all not known or not certain; some or all of the requirements can be changed between iterations	Some known, certain and fixed; some not known or not certain and defined later	Most or all not known or not certain; some or all of the requirements can be changed between iterations
Phasing	Completed once in sequence	Completed once in parallel	Repeated cycle	Repeated cycle	Repeated cycle	Repeated cycle
Product versions	One	One	Several	Several	Several	Several

(Continued)

TABLE 11.2 (Continued)
Life Cycle Comparison Chart

	Waterfall	Parallel	Incremental	Iterative	Incremental and Iterative	Evolutionary
Intermediate form	No	No	Yes	Yes	Yes	Not relevant
Architectural design	Concept architecture designed early in cycle and frozen; design changes very costly	Concept architecture designed early in cycle and frozen; design changes very costly	Core concept architecture (especially interfaces) defined early and frozen; changes can be accepted for architecture that provides functionality in later increments	A high degree of flexibility; all or part of the architecture can be changed at any time	Core concept architecture (especially interfaces) defined early and frozen; changes can be accepted for architecture that provides functionality in later increments	A high degree of flexibility; all or part of the architecture can be changed at any time

project. The critical criteria are those determined by the market: the cycle duration and customer needs and requirements. The cycle duration indicates the durations for which each cycle is most effective. It must be matched to the market expectations on delivery time. For example, if a customer is prepared to wait 2–5 years for the completion of a large complex system, say a power station, then a parallel cycle is appropriate. On the other hand, if a customer is prepared to accept a partial working system, but wants a working system quickly, then an incremental cycle would be appropriate.

Waterfall and parallel cycles perform most effectively if customer requirements are known, certain and fixed (unchanging) during the cycle. The incremental cycle allows some requirements to be deferred until one or more product versions are delivered to the customer. The customer's operational experience can then be used to firm up the requirements for the product versions that are still to be delivered.

The iterative and evolutionary cycles are the most flexible. Customer needs and requirements can be ill defined. As product versions are delivered to the customer, the operational experience can be used to define and redefine requirements. Some or all of the requirements can change between iterations.

Intersecting Life Cycles

Different life cycles have been created for hardware, software and manufacturing system development. One of the major challenges is to effectively harmonise the different cycles. Maier and Rechtin (2002) propose two models. The first harmonises hardware and software cycles, where the hardware can be a product or a manufacturing system, and the second harmonises the product and manufacturing system cycles.

> *Harmonising hardware and software cycles* (Figure 11.9): This can be achieved by using an incremental development life cycle for both hardware and software. Each increment delivers a version of the system, consisting of both hardware and software. The system incremental version is the intersection between the two life cycles. During each increment hardware development will follow a waterfall or parallel cycle. Software development will typically follow an incremental and iterative, or evolutionary, cycle. System versions can be developed solely for in-house use; these are referred to as intermediate forms as illustrated in Figure 11.9. A review will be carried of each version to verify the success of the hardware-software integration.

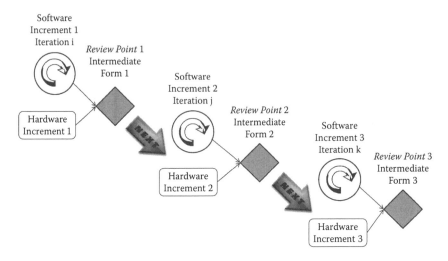

FIGURE 11.9
Harmonising hardware and software cycles.

Harmonising product and manufacturing system cycles: The product and manufacturing system cycles have two phases in common—ramp-up and manufacturing—and this is where the cycles physically intersect, as shown in Figure 11.10. The product life cycle is indicated by the grey shading and the dashed lines. The intersecting phases are indicated by the grey striped boxes. Although the two cycles do not intersect until ramp-up, product and manufacturing system designers should interact before this to ensure the two architectures are properly aligned with each other. The product and manufacturing systems are often jointly designed, as indicated by the dotted line. If prototyping is carried out in both cycles, then it will be another phase of intersection.

Figure 11.10 shows one product cycle intersecting with the manufacturing system cycle. In general, a manufacturing system will make more than one product, so there will be several product cycles intersecting with the manufacturing system cycle, adding to the complexity.

Reviews become even more important when cycles intersect. Figure 11.11 shows an example of the phases and reviews for a Finnish company, Profile Vehicles. The company developed a new architectural design for its products necessitating a major change in the way the company manufactured its products. Three design reviews, 2–4, were created to cover both product design and the design of the manufacturing system. Prototyping through a technology centre and pilot plant trials were critical to the success of this project. Note that there are two reviews to check the manufacturing system readiness (5 and 6).

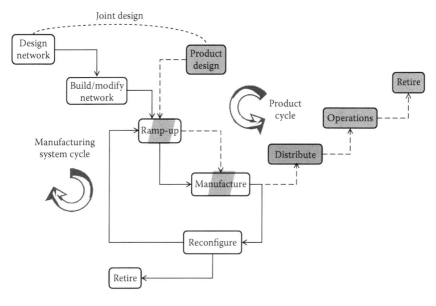

FIGURE 11.10
Harmonising product and manufacturing system cycles.

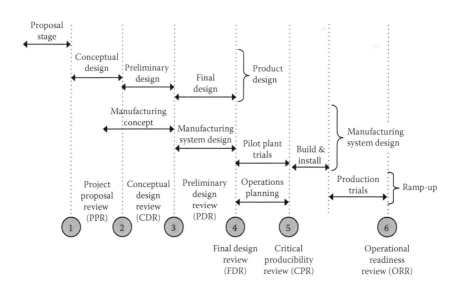

FIGURE 11.11
Product and manufacturing system intersecting cycles.

Specifying Life Cycles and Reviews

The previous sections have shown how to select a life cycle for each type of development effort and how they can be designed to harmonise. This section provides a generic template for defining phases and reviews for the intersecting cycles of Figure 11.11 for batch manufacture. A complete set of phase definitions and reviews is given in Appendix C. If an evolutionary life cycle approach is used, then it is feasible to have only two reviews: one at the beginning to initiate the project and one at the end, an operational readiness review. The design evolution needs to be monitored daily and design decisions made as and when necessary. The decisions must be recorded and monitored as well.

Two general types of reviews are carried out: management and design. A design review focuses solely on the design and design critique. A management review focuses on the business aspects of the development project. The two types of reviews are often carried out separately, but can be combined. The reviews described in our generic template are combined reviews.

The selection of a life cycle, the number of phases in a life cycle, the number of reviews and the documentation to control the design must be useful and easy to use. The review templates are intended to be reasonably comprehensive and serve as a basis for companies to design their own templates. They should not be copied blindly. The reader should simplify and modify the templates to suit his or her specific circumstances.

The phase definition template should have the following format:

- *Goal*: The goal to be achieved by the phase.
- *Description*: A brief description of the phase.
- *Inputs*: The major type of inputs required and their source.
- *Tasks and accountabilities matrix*: This lists the major tasks to be executed in the phase and who is accountable for achieving them.
- *Useful tools*: A list of the main techniques and templates, based on best practices, that can be used during task execution.
- *Deliverables*: A list of the outputs from the phase.

The review template has the following format:

- *Description*: The purpose of the review.
- *Membership*: A list of the members of the review panel, by role.
- *Success criteria*: A list of criteria to be used to determine whether the phase has been completed or not.
- *Documentation*: A written record of the review.

Example: Preliminary Design and Manufacturing Concept Phase

Goal: To generate the product preliminary design integrated with the manufacturing concept.

Description: Design the physical architecture, sub-unit design, design verification plan, and the integration and development of the manufacturing concept.

Inputs:
- Conceptual design phase deliverables.
- Source: Conceptual design review team.

Tasks and accountabilities matrix (Table 11.3):

Useful tools:
- Modelling tools for design (e.g., computer-aided design (CAD), physical mock-ups, etc.)
- Critiquing tools
- Finite element analysis (FEA)
- Failure mode effects analysis (FMEA)
- Worst-case analysis
- Fault tree analysis
- Sneak analysis
- Safety analysis
- Risk management tools
  - Contingency planning
  - De-scope planning
  - Watch lists
  - Critical items and issue list
- Modelling tools for industrial design

TABLE 11.3

Task and Accountabilities Matrix for Preliminary Design and Manufacturing Concept Phase

Task	Team Role Accountable for the Task[a]
Project management	Project manager, team leaders for the product and manufacturing system design teams
Quality management	Chief designers for the product and manufacturing system design teams
Risk management	Team leader and the chief designers
Design management	Chief designers and librarians
Develop marketing plan and sales configuration matrix	Marketing
Design physical product architecture, construct qualification models and prototypes	Chief designer (product), client
Sub-unit design	Designers (product), client
Develop manufacturing concept	Chief designer (manufacturing system), client
Critique	Critics

[a] The role names in the matrix and review team membership are defined in detail in Chapter 16.

- Timescaled Gantt chart
- Checklists and templates
- Flow process charts

Deliverables:

- Detailed industrial design
- Qualification models
- Assembly model or drawings (preliminary design configuration baseline, i.e., product structure)
- Sales configuration matrix
- Preliminary marketing plan
- Sub-unit models or drawings
- Design critique documentation
- Product integration plan
- Product service concept
  - Maintainability
  - Upgradability
- Updated risk plan
- Updated quality plan
- Updated critique plan
- Manufacturing concept
  - Manufacturing functional configuration baseline (i.e., process flow description)
  - Preliminary methods selection
  - Throughput time analysis
    - Time analysis
    - Throughput analysis (production rate)
  - Manufacturing target costing
  - Definition and selection of critical manufacturing equipment
  - Preliminary maintenance concept
  - Preliminary manufacturing quality plan
  - Preliminary manufacturing capital estimate

Example: Preliminary Design and Manufacturing Concept Review

Description: A review to ensure product design can proceed to the final design stage, that the product design is integrated with the manufacturing concept, and that the manufacturing concept can proceed to the manufacturing system design phase. The review assesses whether the product and manufacturing system performance requirements can be achieved within the project budget, schedule, risk and other constraints. It is both a design and management review.

Membership:

- Senior management.
- Executive champion (if appropriate).
- Project manager.
- Client.
- Team leaders.
- Chief designers.

- Chief co-designers.
- Designers.
- Critics (internal and external): External experts should be used to provide independent critique of the design concept.
- Inventors (if appropriate).
- One librarian (to record the meeting).

Success criteria: Success criteria include affirmative answers to the following exit questions:

- Does the current status of the technical effort and design indicate successful production of the product?
- Have the design and manufacturing configuration baselines been established and documented to enable the next phase to proceed with proper configuration management?
- Are the risks known and manageable for the next phase to proceed?
- Can the project be completed within the proposed timeline and budget?
- Are the quality and critique plans adequate for the project to proceed to the next phase?
- Has the marketing plan been established and documented in sufficient detail to proceed to the next phase?

Documentation: Record the review panel decision and get all members of the panel to sign off.

Using Life Cycles for Sustainability Studies

The preceding discussion has focused on using life cycles to improve system development and integrate product design and manufacturing. Life cycle analysis (LCA) can also be used for sustainability studies to assess the environmental impact of products and manufacturing systems. LCA studies vary widely in terms of purpose, the breadth of assessment and the parts of the life cycles that are considered. A full life cycle analysis is called a cradle-to-grave approach. It starts with the extraction of raw materials and ends when a system is retired. If the materials from a system are recycled or remanufactured, then a new cycle begins, starting with the retirement phase of the original system. This is simple to state in concept but difficult to apply in practice. Consider an analysis for a factory to determine its carbon footprint. The study would include a study of its energy usage. But what about the materials it uses? Some of these may be created by supplier and consumer processes that have large carbon footprints. Should these be included, and if so, where do we stop along the supply chain? Do we consider the immediate suppliers and consumers or go further? It is therefore extremely important to define the purpose and scope of a LCA, as these define the boundaries for analysis, especially what will be included and what will not be included.

LCA is normally conducted in four phases (SAIC 2006, ISO 14040). The first phase is definition of the purpose and scope. The purpose defines what the study is expected to achieve, for example, to compare the carbon footprints of two alternative manufacturing processes. The scope defines the breadth and depth of the study. This phase will also provide the background context and define the rules for conducting the study.

The second phase is a life cycle inventory (LCI). It is a data collection phase that produces a list quantifying the amount of pollutants released into the environment, the energy and water usage, and types and amounts of materials consumed by the system being studied.

The third phase is a life cycle impact assessment (LCIA). The items listed in the inventory are assessed in terms of their impacts on human health and environmental ecologies. The purpose of this phase is to show that there is a link between the system being studied and the health of people and the ecological system affected by it. The models are simplified and are suitable only for relative comparisons of risk, not absolute comparisons. They cannot be used for risk assessment.

The final phase is life cycle interpretation. This is a systematic technique to identify, quantify, check and evaluate information from the LCI and LCIA phases and to communicate them effectively (SAIC 2006).

Exercises and Problems

1. Explain how a product life cycle improves the system development process.

2. List and briefly describe the different types of life cycles.

3. Name and briefly describe the different methods for harmonising life cycles.

4. Discuss the advantages and disadvantages of design reviews.

5. The CEO of a small to medium manufacturing company (SMMC) plans to develop a new product. The product development is expected to take 1 year. The technology that will be used is mature and well known. Some customer requirements are known, but there is uncertainty about other critical requirements. The CEO proposes to use a waterfall life cycle model. Is this appropriate? Give reasons for your answer. What model would you propose under these conditions?

References

Alexander, C. 2002. *The nature of order. Book 2: The process of creating life.* Berkeley: Center for Environmental Structure.

Brooks, F.P., Jr. 2010. *The design of design.* Upper Saddle River, NJ: Addison-Wesley.

Clark, K.B., and Wheelwright, S.C. 1993. *Managing new product and process development.* New York: Free Press.

ISO 14040. 2006. *Environmental management—Life cycle assessment—Principles and framework.* International Organisation for Standardisation.

Maier, M.W., and Rechtin, E. 2002. *The art of systems architecting.* 2nd ed. New York: CRC Press.

SAIC. 2006. *Life cycle assessment: Principles and practice.* EPA/600/R-06/060. Cincinnati, OH: EPA.

Shisko, R. 1995. *NASA systems engineering handbook.* Washington, DC: NASA.

Shore, J., and Warden, S. 2007. *The art of agile development.* Sebastopol, CA: O'Reilly.

12

Product and Manufacturing System Architecture Integration

Introduction

Chapter 4 explained how manufacturing systems can be designed for cellular manufacturing given a pre-defined family of products. The product designs were taken as fixed. Here we break that assumption and address the issue of how to jointly design products and manufacturing systems for cellular manufacturing.

The concept map for the chapter is shown in Figure 12.1.

The integration of product and manufacturing system architectures is achieved through a product platform architecture and a manufacturing system designed to match the product architecture. A modular product family architecture (MPFA) is suitable for batch design and batch manufacturing. Common modules make up the platform, and design variants are produced by adding or substituting variant modules.

The ideal manufacturing system is designed to match a modular product architecture. It consists of basic manufacturing units (BMUs) assembled into factories. A BMU is a self-controlled cell containing technological equipment and people.

Product Platforms

Many companies design their products separately to meet different market requirements, with little or no sharing of components between products. The products compete against each other for development funds. During the selection process a portfolio of products is chosen for development. The disadvantages of this approach compared to a product platform approach are longer development lead times and higher development, manufacturing and service costs.

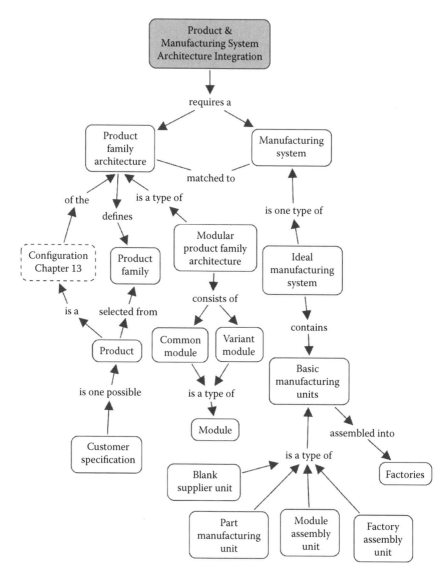

FIGURE 12.1
Concept map for Chapter 12.

A product platform approach is based on designing, at the same time, a set of different products to meet different markets. The aim is to reduce the number of components and sub-system variants by sharing common components and sub-systems amongst the different products. The common components and sub-systems form a *platform* from which derivative products can be developed and made (Meyer and Lehnard 1997). The set of products that share the same platform is called a product family. Different market requirements are met by adding customised components and sub-systems to the product platform. Some examples of products based on product platforms are Volkswagen (same platform used for Volkswagen, Audi, Seat and Skoda), Sony Walkman, HP ink jet and laser jet printers, Black and Decker power tools, Boeing 747, and Rolls Royce RB211/Trent engines.

A product platform is the basis for developing product variants in the short term and for extension in the long term. The platform architecture for product variants and for the extended platform is the same (same number and types of sub-systems and interfaces). Platform extension involves technological upgrading of one or more sub-systems or modules. The platform must be designed with an open architecture so that the variants and extension can be carried out without much effort. This in turn requires clearly defined standard interfaces.

There are two general approaches for designing product platforms: scale-based and modular. In the scale-based approach the platform extensions and variants are produced by changing certain design parameters. Platform architectural design involves defining the parameters that will not change, defining the scaling parameters, specifying the common sub-systems and their interfaces, and integrating the design with manufacturing. Scale-based product family architectures are common in the automotive and aerospace industries.

The modular approach divides a product into modules. A modular architecture consists of common and variant modules. A common set of modules (the platform) is contained in all products in the family. Variant modules are added and substituted to give customer variants.

Any product can be designed for modularisation, but high-performance products will normally be excluded because a modular design will have lower performance than a high-performance design. The performance degradation will outweigh the benefits of modularity.

One major advantage of a modular product architecture is it allows the use of an assemble-to-order delivery process, drastically cutting order lead times to customers, and hence giving a small to medium manufacturing company (SMMC) a market advantage.

Definitions of key concepts for this chapter are given below:

Product family: A set of products or product variants that share some common components or sub-systems but also contain different components and sub-systems to meet different customer requirements.

Product platform: The set of common components/sub-systems that is contained in all products in a product family.

The variants in a product family can be produced two ways: by scaling up or down using scaling variables or by the substitution/addition of modules.

Scale-based product family: A product family in which product variants are formed through different values of scaling variables.

Scale-based product family architecture: The platform and scaling design features and their interrelationships.

Module: A set of parts that are designed to achieve a specific function. A module can consist of one part only. It can be an assembly but need not be. Modules must have clearly defined interfaces.

Modular product family: A product family in which product variants are formed by the substitution/addition of modules.

Modular product family architecture (MPFA): The common platform modules and their features, the variant modules and their features, and the interrelationships between these.

Modularisation: The decomposition of a product into modules with clearly specified interfaces.

The rest of this chapter discusses modular product architectures and how they can be integrated with a manufacturing system.

Modular Product Family Architectures

A modular product family architecture (MPFA) is illustrated in Figure 12.2. As can be seen, working from right to left, the final product is divided into modules. The modules are created from parts made at the part level. All parts

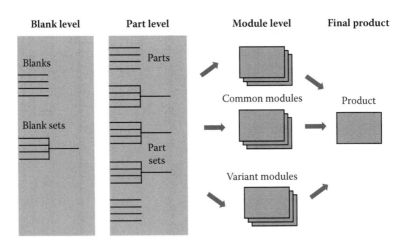

FIGURE 12.2
Illustrative modular product family architecture. (Adapted from Lapinleimu, I., Ideal Factory. Theory of Factory Planning, Produceability and Ideality, PhD thesis, Institute of Production Engineering, Tampere University of Technology, 2001.)

should be made to match the exact numbers required for each module. These are called part sets. Efficient batch manufacturing requires parts to be made in part sets, as discussed in Chapter 4. Finally, the parts themselves are made from material blanks. Blank and part manufacturing are separated because their manufacturing technologies are different. The blank and part levels are not part of the modular architecture, but they feed into it.

There is a big difference between a sub-assembly and a module. Sub-assemblies are created because a design does not allow cost-effective assembly directly from parts. Poor designs will have many sub-assemblies. On the other hand, a modular design requires only two types of assembly processes: one for the modules and one for the final assembly. The advantage of modular design is that modules can be assembled concurrently, reducing the manufacturing lead time. The modules must be designed so they can be assembled quickly to create the final product.

There are a number of methods for developing modules. Most of them focus on design only. But the product architecture affects a wide range of stakeholders, for example, manufacturing, sales, maintenance and service, senior management (in terms of company strategy), and the environment. Ericsson and Erixon (1999) have developed an excellent method that takes into account different stakeholder viewpoints. The basic concept behind the method is to use what they call *module drivers* to identify modules. The drivers ensure that different types of variation are limited to one or a few modules. In other words, product variety is achieved through localised variation.

Ericsson and Erixon (E&E) identified 12 module drivers[TM]:

- Carryover: Parts that are not likely to change in the future and therefore can be "carried over" to new products.
- Technology evolution: Parts that are likely to change due to changes in technology or customer requirements.
- Planned product changes: Parts that are planned for change.
- Different specification: Parts that vary significantly and which are bundled together to minimise impact on the architecture.
- Styling: Parts that change due to fashion and customer taste.
- Common unit: Parts with functions required by all products in the family.
- Process or organisation: Parts bundled together for more efficient manufacturing.
- Separate testing: Parts that are bundled to allow separate testing of modules before delivery to final assembly.
- Available from supplier: Standard modules purchased from suppliers.

- Service and maintenance: Parts requiring frequent service and maintenance are bundled together.
- Upgrading: Parts designed to allow a module to be upgraded.
- Recycling: Parts designed for recycling or remanufacture.

The authors have refined E&E's schema for identifying modules. We do not use the term *driver*, as it is not appropriate in our view. Instead, our schema defines generic classes. The criteria for the classes are:

- Viewpoint: The stakeholder who benefits most from the module. The stakeholders are strategic management, design, customer, manufacturing, maintenance and operations support, and environment.
- Basis: Key criteria for identifying characteristics to support the viewpoint.
- Type of variation that is controlled.
- The factor being controlled (cost, quality, etc.).
- Design goal.

Our schema results in six module classes (Table 12.1) that are identified by stakeholder.

Module definitions by other researchers such as Ulrich and Tung (1991) and Ulrich and Eppinger (2000) relate to our schema as follows. The design class includes component sharing modularity (Ulrich and Tung 1991), where a design component is the common module, and component swapping modularity (Ulrich and Tung 1991), where the base unit is the common module. "Interface modules"—bus and sectional modularity (Ulrich and Eppinger 2000)—are also included in the design class.

Our schema maps to E&E as shown in Table 12.2.

The advantages of our class definitions compared to E&E are:

- There are fewer classes so they are easier to apply.
- We have more clearly identified the characteristics of the classes so they are easier to understand and apply.
- We have identified other types of modules, such as packaging modules (volume and lifting modules). These modules are required for very large products that must be broken down to suit the logistic transportation requirements, e.g., container size and maximum lifting capacity of cranes.

Modules can be assembled together by stacking (horizontally or vertically) or by attaching to a base unit (Ericsson and Erixon 1999), as illustrated in Figure 12.3. Aircraft jet engines are usually assembled by stacking. Vertical stacking is common for small engines. Rolls Royce Trent engines are assembled using horizontal stacking.

TABLE 12.1

Module Class Definitions

Viewpoint	Basis—Identifying Criteria and Characteristics	Type of Variation Controlled	Controlled Factor	Design Goal
Strategic management	Planned evolution of product family (expected markets and technological evolution)	Infrequent, some changes are forecast, but others are unknown Step changes	Product change	Minimise effect of technology and product evolution changes on the architecture Design for long product lifetime
Design	Reuse of part sub-sets that perform common functions Often these parts are invisible	None—no variation between products, same parts for all	Cost Reliability	Maximise size of platform compared to variants to reduce costs and ensure reliability
Customer	Customer requirements Aesthetic and user interaction modules are visible	Very frequent and known	Customer performance requirements Customer aesthetic requirements	Minimise effect of changes in customer aesthetic and functional and performance requirements on the architecture
Manufacturing	Matching product structure to manufacturing network—processing and material handling system constraints	None or very little. If variation is allowed, it will be based on some scaling parameter or on process similarity (group technology)	Cost Lead time	Minimise the impact of the architecture on manufacturing Minimise testing costs Minimise development costs by using supplier modules Minimise the impact of the architecture on the logistic network through packaging modules (volume modules and lifting modules)

(Continued)

TABLE 12.1 (*Continued*)

Module Class Definitions

Viewpoint	Basis—Identifying Criteria and Characteristics	Type of Variation Controlled	Controlled Factor	Design Goal
Maintenance	Providing cost-effective maintenance and operations support There may be a special module for customer testing or verification of system status	None or some variation is frequent if it exists	Cost Availability	Minimise maintenance and service costs Maximise product availability
Environment	Minimising adverse impact on or enhancing the environment Includes product packaging as a module for reuse or recycle	None or some variation is frequent if it exists	Company reputation Environmental features	Minimise adverse effect on the environment and if possible enhance the environment Ensure compliance with regulatory authorities

TABLE 12.2

Mapping between Authors' and E&E's Schemas

Authors' Stakeholder Classes	E&E's Drivers™
Strategic management	Technology evolution
	Planned product change
	Upgrading
Design	Reuse
	Carryover
Customer	Styling
	Different specification
Manufacturing	Process or organisation
	Supplier available
Maintenance	Separate testing
	Service and maintenance
Environment	Recycling

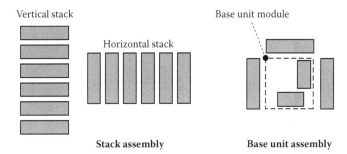

FIGURE 12.3
Generic ways to assemble modules.

Modules may be attached on the outside, inside, or both, of the base unit in the base unit assembly. An example of both is an ambulance where the body is the base unit. Some parts, e.g., the flashing lights on the roof, are mounted on the exterior. Other modules, such as seats and medical equipment, are mounted in the interior.

Configuration

MPFA contains all the possible modules that can be included in a particular customer specification. Each customer selects the required modules from this pre-defined set to create a customer-specified product, which is one configuration out of all the possible configurations that can be created by selecting and combining modules. *A modular product family architecture is a configurable product.* However, the reverse is not true. It is possible to have a configurable product without a platform.

A configuration model defines a set of rules that can generate all possible variants of the product family and creates a particular variant from a customer order specification. The model can be implemented manually or automatically.

Module and configuration model development are two sides of the same coin. Module development creates modules from product variety. The configurable model generates product variety from modules. Chapter 13 discusses configuration in detail.

MPFA Generating Principles

There are two types of system architectures: functional and structural (Chapter 2). A MPFA is generated from these architectures as shown in Figure 12.4. The platform modules are generated from the structural architecture. They are structural modules. Structural modules should be designed to map directly onto the manufacturing system architecture, as this minimises costs.

The variant modules are generated from the functional architecture. They are functional modules. The functional architecture is derived by decomposition of a product family's functions in a top-down manner. The architecture is then iteratively partitioned and perhaps re-decomposed until a satisfactory set of modules is obtained. The aim is to have a set of functional

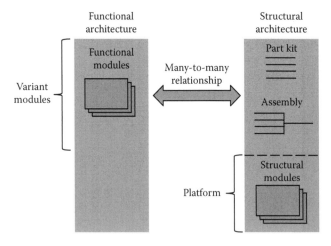

FIGURE 12.4
MPFA generating principles.

modules that can be easily generated from product features and their options and are capable of generating all possible product variants.

The functional modules are created from the structural architecture as part kits or assemblies. The relationship between the functional and structural architectures is many to many. Hence there are many ways to create part kits and assemblies to form functional modules. In general, the functional modules will not map directly onto the manufacturing system architecture, but with careful design the mismatch can be minimised.

A configuration matrix generates functional modules and the platform structural modules. The part kits and assemblies for functional modules are generated by rules and routes in a materials requirements planning (MRP) or enterprise resource planning (ERP) system.

The Ideal Manufacturing System

The ideal manufacturing system is one whose architecture matches the MPFA. The system is created from basic manufacturing units (BMUs). The BMUs are semi-autonomous cells (Figure 12.5). Each cell consists of technological equipment and people and is self-controlled as far as possible. The purpose of a BMU is to transform materials into parts and parts into assemblies. The primary inputs are parts and materials and the primary outputs are parts and assemblies. The primary inputs and outputs are the basis for identifying BMUs.

Other inputs required for the unit to operate include energy, information, etc. Other outputs include waste materials. Operating constraints imposed

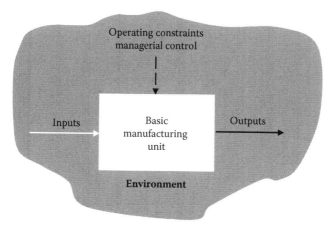

FIGURE 12.5
Basic manufacturing unit.

by the environment and managerial commands and orders are indicated separately to distinguish them from the material flow. They are indicated by a dashed line in the figure.

There are four different types of BMU:

- Blank supplier unit (BSU) that supplies blanks of material ready for further processing. The blanks may be supplied as individual parts or in matched sets (blank sets).
- Part manufacturing unit (PMU) that makes parts. The parts may be supplied as individual parts or in matched sets (part sets or sub-assemblies that are not modules).
- Module assembly unit (MAU) that collect parts into sets or assembles them as modules.
- Final assembly unit (FAU) that assembles modules into final products.

BMUs are identified separately from PMUs because blanking manufacturing technology is different from part manufacturing technology.

Each unit will make several products. If production volumes are large enough, then a unit may make one product.

The ideal material flow in a manufacturing network is one in which materials flow directly from blank manufacturing to an end product without cross-flows or backtracking (Figure 12.6). The logic is the same as for group

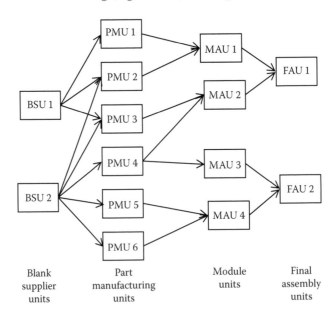

FIGURE 12.6
Ideal manufacturing system—a network of BMUs.

TABLE 12.3

Production Planning and Control Focus

Unit	Production Planning And Control Focus
BSU	Service and stock oriented
PMU	Technology oriented
MAU	Product oriented
FAU	Customer oriented

technology. Ideal flow makes networking between units easier, especially if the units are owned independently, as in Finland. Note the similarity between Figures 12.2 and 12.6. This is essential for mapping the modular product architecture directly onto the manufacturing network and vice versa.

The ideal flow depends on the products being designed with a modular architecture. The network architecture is designed to match the MPFA. Modules may be assemblies, but this is not essential. For example, a styling module may consist of several parts that are not directly physically connected. "Assembly" consists of accumulating the parts and packaging them in a kit as a matched part set. A MAU that operates in this manner is acting as a marshalling store.

Production planning and control focus is different for the four different types of units, as shown in Table 12.3. The BMUs can be combined in different ways to form factories, as illustrated in Figure 12.7. Some factories can be dedicated to part manufacturing only, some to assembling modules only, and others can make parts and assemble modules. The formation of factories is a secondary consideration.

Illustrative Modular Design Example

Three young graduates decide to set up their own company to make configurable classroom table furniture for primary schools. The design concept was inspired by bees. Bee honeycombs have a hexagonal lattice structure. This has been shown to be a very efficient structure for filling space. Thus the concept of hexagonal tables was born. The tables will be mounted on wheels so they can be reconfigured easily and quickly. Some sample table arrangements are shown in Figure 12.8.

The team gave some thought to the tabletop colours. They decided on three colour schemes: day and night, sea, and sky, rather than allowing colours to be mixed randomly. They also decided that the legs and wheels would be the same for all colour schemes. The wheels would be detachable because they are high-wear components and require frequent changing. They chose

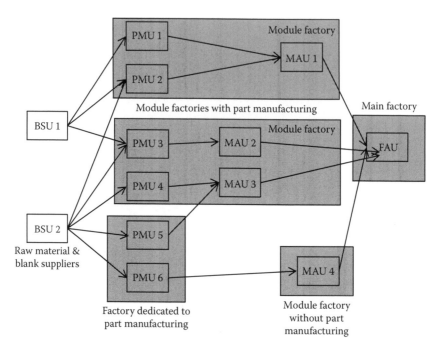

FIGURE 12.7
Formation of factories—some possibilities. (Adapted from Lapinleimu, I., Ideal Factory. Theory of Factory Planning, Produceability and Ideality, PhD thesis, Institute of Production Engineering, Tampere University of Technology, 2001.)

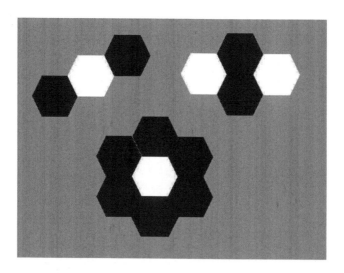

FIGURE 12.8
Possible table arrangements.

a neutral colour grey for these components: light grey for the legs and dark grey for the wheels. The modular architecture is shown in Table 12.4.

The legs belong to the strategic management class because the team believed that it could develop other markets, such as secondary schools and pre-schools. Products for these markets would be developed by simply lengthening or shortening the table legs.

The next consideration was deciding the colours for each colour scheme. The team's first choice is shown in Table 12.5.

The platform analysis for this choice is shown in Table 12.6. The analysis simply compares the number of modules in the platform to the number of variant modules. A larger platform reduces costs and improves reliability, but reduces product variety, and vice versa.

The team was not happy with this first choice, as the platform percentage was too low. So they decided to use white tops to replace the light green and blue ones. The colour scheme is shown in Figure 12.9A (dark grey represents green and light grey represents dark blue).

The platform analysis is shown in Table 12.7. The platform has three modules now because the white tops are also common.

TABLE 12.4

Initial Modular Architecture

Module Variants	Module Class
Tabletop	Customer (styling module)
Platform	
Leg	Strategic management
Wheel	Maintenance

TABLE 12.5

First Choice of Colours

Colour Scheme	Colours
Day and night	Black and white
Sea	Dark green and light green
Sky	Dark blue and light blue

TABLE 12.6

Platform Analysis for First Choice

	No. of Modules	%
Variants	6	75
Platform	2	25

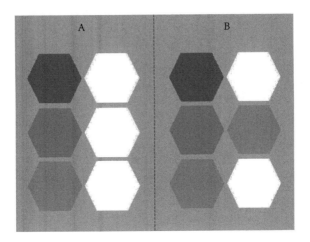

FIGURE 12.9
Two possible colour schemes.

TABLE 12.7

Platform Analysis for Option 12.9A

	No. of Modules	%
Variants	3	50
Platform	3	50

TABLE 12.8

Platform Analysis for Option 12.9B

	No. of Modules	%
Variants	4	67
Platform	2	33

The team felt this was not a good option as there was too little product variety. They finally settled on the colour scheme shown in Figure 12.9B. The sea colour scheme was changed to dark blue (represented by dark grey) and light blue (shown as light grey). The light blue is the same colour as for the sky colour scheme. The platform analysis is shown in Table 12.8.

The example highlights a very important point. Modules that are common across some, but not all, products help to reduce module variation. In option 12.8B, white and light blue tops are common for different sets of products. It is important to remember this when designing modules and not focus solely on the product platform.

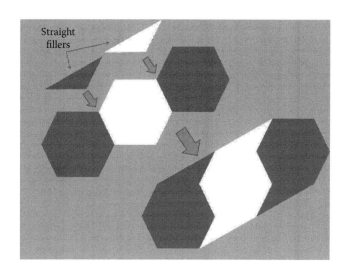

FIGURE 12.10
Straight fillers.

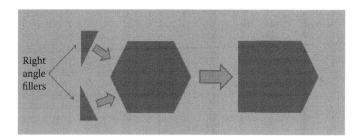

FIGURE 12.11
Right-angle fillers.

The final part of the design concept was to design fillers. Straight fillers are used to join tables together and provide a straight edge (Figure 12.10). Right-angle fillers are used to produce tables that can fit neatly into corners or to produce a square table (Figure 12.11). If a corner table is adjoined to another table, then a straight filler is required to fill the gap between the wall and the table. Fillers are an option.

The MPFA is summarised in Table 12.9 and the platform analysis in Table 12.10. Note how the addition of fillers has reduced the platform percentage. Each additional feature and its options generate product variety and reduce the percentage contribution of the platform.

The principles for designing modular architectures are relatively straight-forward. Their application to real products can be very difficult, as will be demonstrated in Chapter 13 in the case study on modular design of forestry equipment.

TABLE 12.9

MFPA for Configurable Tables

Product Feature	Product Feature Options	Modules	Module Class
Module Variants			
Colour scheme	Day and night	Black top	Customer (styling)
		White top	Customer (styling), design (reuse)
	Sea	Dark blue top	Customer (styling)
		Light blue top	Customer (styling), design (reuse)
	Sky	Light blue top	Customer (styling), design (reuse)
		White top	Customer (styling), design (reuse)
Filler	Straight	Black straight filler	Customer
		White straight filler	Customer
		Dark blue straight filler	Customer
		Light blue straight filler	Customer
	Right angle	Black right-angle filler	Customer
		White right-angle filler	Customer
		Dark blue right-angle filler	Customer
		Light blue right-angle filler	Customer
		Platform	
		Leg	Strategic management
		Wheel	Maintenance

TABLE 12.10

Platform Analysis for Table 12.9

	No. of Modules	%
Variants	12	86
Platform	2	14

Design Integration Process

Figure 11.10 showed how the product and manufacturing system life cycles intersect. Joint design of the product and manufacturing system was shown as a dotted line. Figure 11.11 showed how the joint design processes can be harmonised. In this section we briefly explain the main design activities performed during joint design. The joint design process can be divided into three steps:

Step 1: Analysis

Step 2: Conceptual product design and manufacturing concept

Step 3: Embodiment design and preliminary manufacturing system design

Analysis

The goal of this step is to generate the product development project proposal and to obtain approval to proceed with the project. The main activities are market research and analysis, business context analysis using EASTEEP, and strategic technology and product development planning. Market research aims at determining the likely markets for the SMMC now and in the future, discovering the type of products and their features that customers want now and in the future, and a competitor analysis. The amount of effort required to carry out these activities can be considerably reduced by focusing on one or two lead customers (Chapter 3).

Business context analysis aims at identifying the key characteristics of the business environment that will have an impact over the product life cycle. The authors' EASTEEP framework (Section II) can be used to guide the analysis.

The outputs of the two analyses are inputs to management to develop plans for technology and product evolution. The management plans will determine whether a product structure analysis will be carried out or not. If management decides to change from a non-modular to a modular product architecture, then an analysis is not necessary because the product will be re-designed. If the current products are partly modular or there is an existing product platform, then an analysis may be required.

This step ends with a project proposal review.

Conceptual Product Design and Manufacturing Concept

The goal of this step is to develop the product and manufacturing system functional architectures. The activities in this step include:

- Develop the product functional architecture. The product features and options are determined, the functional modules are designed and mapped to the features and options, and the platform structural modules are determined. A configuration matrix should be developed to formally record and verify the architecture.

- Demonstrate technical and manufacturing feasibility, preferably with a physical prototype. A technology development centre is useful for designing and testing prototypes.
- Prepare a preliminary manufacturing plan (rough-level make and buy analysis) and functional process flowcharts.
- Prepare preliminary risk, quality and project plans.
- Prepare a critique plan.

The step ends with a conceptual design review.

Embodiment Design and Preliminary Manufacturing System Design

The activities in this step include:

- Design the product structural architecture for the functional modules and refine the platform modules.
- Develop an initial manufacturing system (supply chain network) and map to the product architecture.
- Conduct a throughput time analysis (Appendix D explains how to do this).
- Prepare costing targets for manufacturing.
- Define and select the type of assembly process and critical manufacturing equipment.
- Prepare the maintenance concept.

The step ends with a preliminary design review.

Exercises and Problems

1. Describe a product platform.
2. Is it possible to have a product platform that is not modular? Explain your answer.
3. Is it possible to have a configurable product without a platform? Explain your answer.

4. Explain the relationship between a product configuration and a product family.

5. Briefly describe the ideal manufacturing system. How does the MPFA map onto it?

References

Ericsson, A., and Erixon, G. 1999. *Controlling design variants: Modular product platforms.* Dearborn, MI: SME.

Lapinleimu, I. 2001. Ideal factory. Theory of factory planning, produceability and ideality. PhD thesis, Institute of Production Engineering, Tampere University of Technology.

Meyer, M.H., and Lehnard, A.P. 1997. *The power of product platforms.* New York: Free Press.

Ulrich, K., and Eppinger, S. 2000. *Product design and development.* New York: McGraw-Hill.

Ulrich, K., and Tung, K. 1991. Fundamentals of product modularity. In *Proceedings of the ASME Winter Annual Meeting Symposium on Issues in Design/Manufacturing Integration*, Atlanta, GA, pp. 73–79.

13

Modular Configuration

This chapter is based on Juho Nummela's (2004) thesis. Juho is the president and CEO of Ponsse, a Finnish company that makes forestry machines and equipment. Juho developed an automatic, modular configurator for Ponsse during his research studies. The authors thank and acknowledge Juho for permission to reprint parts of his thesis and photographs of Ponsse equipment.

A configuration is one product from all possible products that can be selected from a product family architecture. A configuration model defines a set of rules that can generate all possible variants of the product family and creates a particular variant from a customer order specification. The rules act as constraints to prevent the configuration model from creating invalid variants.

The configuration model must capture all the interrelationships between the product features a customer can select. It must also allow for different sequences of feature selection by customers and sale staff. Configuration matrices are an effective way to represent and maintain configuration knowledge.

Terms used in this chapter are defined below:

Configuration: A product platform and a set of variant modules to meet a particular customer order specification. It is one selection out of all possible selections.

Configuration model: The representation of product features and modules and their interrelationships that can generate valid configurations.

Configuration matrix: A matrix that implements the configuration model. It is a component of the configurator.

Configurator: A tool that supports configuration management. A configurator does not have to be based on a configuration matrix and can be manually operated or automated.

Configuration Matrices

Configuration matrices are designed to help people in a small to medium manufacturing company (SMMC) understand product architectures, to reveal the usually hidden knowledge that is critical in order to effectively produce configurations, and to maintain the configuration knowledge. Matrices can present the knowledge to be used for both manually and automatically controlled configuration processes as they are formed to support the configuration task generally. The main purpose for the configuration matrices is to present all the options saleable at the moment and the connections between the combination of the saleable options and the modules for production. The configuration rules are established by revealing the configuration knowledge, much of which is implicit.

The matrices are square matrices (Table 13.1). The cells heading the rows and columns are the modules and their features, including the options. The features are indicated by numbers and the options by letters. Thus feature 1 has two options, A and B. Modules are indicated by numbers.

Variant modules that will be selected by a set of options are located directly below those options. Thus the row for module 1 starts after the row for option 1B.

The platform modules form the bottom and far right parts of the matrix.

A dependency between options and modules is indicated by a cross in the matrix. There are no entries along the diagonal. The matrix is symmetric—dependencies are entered above and below the diagonal. The reason for this is that the matrix has a particular row and column sequence. Features can be selected in any order, and therefore the matrix must allow for this. Dependencies can be read from column to row or row to column.

The dependencies are defined by rules using AND and OR operators. Marketing rules control the interdependencies between saleable (feature) options. For example, in Table 13.2 the marketing rules are indicated in the cell intersections between 1A and 1B; 2A, 2B and 2C; (1A and 1B) and (3A and 3B). The latter indicates dependencies between features 1 and 3.

Manufacturing rules control the interdependencies between options and modules. The rules are indicated in the cells where the feature options intersect the module rows/columns. A rule may contain a single option that is sufficient for selecting modules or a set of options in combination. A customer-specified product is obtained when a customer has chosen all desired features and saleable options. The matrix automatically generates the modules for the customer's specification.

Blank spaces have two different interpretations. If the cells for all options for one feature have no relationship to another feature, then the cells are blanks. In Table 13.2 feature 2 has no relationship with feature 3. On the other hand, if there is a dependency between two features, then a blank space means that the option is not allowable. This is the situation for features 1 and 3. The possible choices are shown in Table 13.3. Thus if option 1A is chosen, option 3A is not possible and vice versa.

Module selection depends on the feature options that have already been selected. Assume that options 1A and 2B have been selected. Then selecting 3A will lead to the combinations shown in Table 13.4 column 2.

TABLE 13.1

Illustrative Configuration Matrix

	Option 1A	Option 1B	Module 1	Module 2	Module 2	...	...	...	...	Platform	Module P1	Module P2	Module P3	Module Pn
Option 1A														
Option 1B	×			×	×									
Module 1	×			×	×									
Module 2		×	×		×									
Module 3		×	×	×										
...														
...														
...														
...														
Platform														
Module P1														
Module P2														
Module P3														
Module Pn														

Source: Nummela, J., Integrated Configuration Knowledge Management by Configuration Matrices—A Framework for Representing Configuration Knowledge to Organization, PhD thesis, Institute of Production Engineering, Tampere University of Technology, 2004. With permission.

TABLE 13.2

Illustrative Variant Portion of a Configuration Matrix

	Option 1A	Option 1B	Module 1	Module 2	Module 3	Option 2A	Option 2B	Option 2C	Module 4	Module 5	Option 3A	Option 3B	Module 6	Module 7	Module 8	Module 9	Module 10
Option 1A	x	x									x	x	x	x	x	x	x
Option 1B	x		x	x	x						x	x		x		x	x
Module 1				x	x												
Module 2		x	x														
Module 3		x	x	x													
Option 2A							x	x	x								
Option 2B						x		x		x							
Option 2C						x	x		x								
Module 4						x		x		x							
Module 5							x		x								
Option 3A	x	x									x	x	x	x	x	x	
Option 3B	x	x											x				
Module 6		x									x			x	x	x	x
Module 7	x										x		x		x	x	x
Module 8		x									x		x	x		x	x
Module 9	x											x	x	x	x		x
Module 10												x	x	x	x	x	

Source: Nummela, J., Integrated Configuration Knowledge Management by Configuration Matrices—A Framework for Representing Configuration Knowledge to Organization, PhD thesis, Institute of Production Engineering, Tampere University of Technology, 2004. With permission.

TABLE 13.3

Possible Choices for Features 1 and 3

	Option 1A	Option 1B
Option 3A	Not possible	Possible
Option 3B	Possible	Possible

Source: Nummela, J., Integrated Configuration Knowledge Management by Configuration Matrices—A Framework for Representing Configuration Knowledge to Organization, PhD thesis, Institute of Production Engineering, Tampere University of Technology, 2004. With permission.

TABLE 13.4

Possible Module Selection for Example

	Selectable for Selection String 1A AND 2B AND 3A	Selection Strings That Will Select the Module
Module 6	No	3A AND 1B
Module 7	Yes	3A AND 1A
Module 8	No	3A AND 1B
Module 9	Yes	3A AND 1A
Module 10	Yes	3A AND (1A OR 1B)

Source: Nummela, J., Integrated Configuration Knowledge Management by Configuration Matrices—A Framework for Representing Configuration Knowledge to Organization, PhD thesis, Institute of Production Engineering, Tampere University of Technology, 2004. With permission.

Column 3 shows the selection string required to select that particular module. Note that feature 2 has no effect on the selection.

Configuration matrices link product development, manufacturing and marketing/sales, as shown in Figure 13.1. If the matrices are developed concurrently with the development of the modular product architecture, then they provide a powerful means to involve these three major stakeholders in the design process and to document expert knowledge about the product and how it can be made.

Process for Creating Configuration Matrices

Configuration matrices can be generated for both new and existing products. As product design is responsible for generating configuration knowledge for the rest of the SMMC, a process is needed to tie together the configuration matrices and product design.

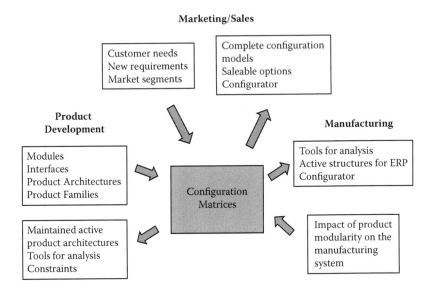

FIGURE 13.1
Function cross-linking through interaction matrices. (From Nummela, J., Integrated Configuration Knowledge Management by Configuration Matrices—A Framework for Representing Configuration Knowledge to Organization, PhD Thesis, Institute of Production Engineering, Tampere University of Technology, 2004. With permission.)

The development of a configuration matrix should be closely tied to the design integration process described in Chapter 12. A matrix can be developed in three stages. The first stage is to develop a configuration plan. Planning can start when customer requirements and management product evolution strategy are confirmed. The initial plan for configurability can be considered to be the basis for configuration matrices as variety, commonality, differentiation, outsourcing and upgradeability are considered. The plan will specify the kind of information required and the source.

The next stage is to develop the initial matrix. This should occur as early as possible in the design process. The initial matrix can be created once the first modular product architecture has been established and there is a clear understanding and consensus of what the saleable options of the product will be. The following steps can be used to create the initial matrix:

1. Define the generic list of saleable options.
2. Generate lists of:
   a. Common modules
   b. Variant modules
3. Define the sequence of the saleable options (row and column order). Only one sequence can be selected, but all sequences will be built in to the matrix.

4. Generate the square matrix by listing the options and modules according to the sequence.
5. Define the dependencies of the modules with respect to the options.
6. Finalise the sequence of the matrix.

The initial matrix can be used to verify the initial modular product platform.

The third stage is to update and maintain the matrix as the product platform evolves. The configuration matrices can be used to effectively consider changes during the design phase by analysing the dependencies in the matrices. The emphasis should be targeted on the platform versus the variable part of product since there is cost for providing more variety; a larger platform with respect to the variants is more cost-effective.

After the product is released to customers the matrix will be used to generate shop orders for manufacturing. Furthermore, the matrix can be updated to incorporate product upgrades, product extension and changes in the manufacturing system.

Configuration Maintenance

After the configuration knowledge has been documented it needs to be maintained. As time goes by and the life cycle of a product family advances, changes to the product structure and to the configuration knowledge are likely to take place. Well-established change management processes are essential to control and maintain the configuration knowledge; i.e., the changes from product design must be systematically handled in order to avoid inconsistencies in configuration models.

Configuration matrices offer a visual tool to handle the changing configuration knowledge. In the case of configurable products the configuration knowledge needs to be valid at all times regardless of the configuration process; i.e., the configuration knowledge needs to be valid for both manual and automatic configuration processes at all times. There are many possibilities for organising configurator maintenance; i.e., it can be the responsibility of manufacturing, product design or a configurator maintenance support team. One of the main reasons to present configuration knowledge with matrices is the easy management of configuration rules in the changing environments. If the rate of change is frequent enough, the management of configuration rules in configurators can become impossible mainly because of the dependencies between saleable options. Tools and features that configuration matrices offer for configurator maintenance are:

- Interface to maintain the configuration matrices:
  - Product architecture

- • Manufacturing configuration rules
- • Marketing configuration rules
- Integration with configurator
- Visual presentation of configuration knowledge
- Documentation of product architectures
- Documentation of generic feature structures
- Documentation of platforms
- Documentation of variants

Configuration matrices can be built manually by using Excel sheets or by using a software system. A software system helps the user to build configuration matrices and confirms that the matrices are built systematically the same way in order for other software systems to interpret the matrices correctly. The same software system can be used to maintain the matrices as well.

Integration with the configurator is of great importance if the frequency of change is relatively high; i.e., all the features and elements needed to set up the configurator would automatically be loaded from the configuration matrices into the configurator. In addition, the configuration knowledge is changing as product development makes changes to products. This implies the need for versioning the matrices in order to secure the existence of configuration knowledge through the entire life cycle of a product family. The matrix versions must match the product versions.

Case Study: Ponsse

The case study describes how Ponsse changed from using a manual configurator to an automatic configurator based on configuration matrices. The manual configurator generates functional product modules. The functional modules are, however, not suitable for manufacturing. Manufacturing cannot make and build functions. The configuration matrices convert the functional modular structure of the products to structures and bills of materials suitable for manufacturing.

The Problem

Ponsse manufactures and markets forest machines for the cut-to-length method of mechanised logging as well as wood harvesting-related information technology. The Ponsse Group consists of the parent company Ponsse Oyj in Vieremä, Finland, and subsidiaries in Brazil, China, France, Norway,

Russia, Sweden, the UK and the United States. Ponsse is the second largest forestry machine supplier in the world. It is truly a hidden champion.

Machines produced by Ponsse can be divided into harvesters, forwarders, harwarders, cranes and harvester heads. Harvesters take care of the cut-to-length method, forwarders collect the harvested wood and harwarders are combination machines able to perform both tasks. Ponsse has four types of harvester machines: Ergo, Beaver, Bear and Fox. The forwarders are Buffalo, Buffalo King, Gazelle, Elk, Elephant and Elephant King. The product family of harwarders is Buffalo Dual. The main products are shown in Figure 13.2: left to right are a Harvester, Forwarder and Harwarder. A detailed view of a Harvester, an eight-wheel Ergo, is shown in Figure 13.3. Figure 13.4 shows the Ergo (harvester) and Elk (forwarder) in action. The Elk is on the left.

One of the most challenging product groups is harvester heads. Figure 13.5 presents the different harvester heads available at the time of the research. From the upper left corner the first product is the smallest harvester head H53. Then the following four pictures present the harvester head family 60 with models H60, H60E, H60BW and HW60. Finally, the last picture in the lower right-hand corner presents the biggest harvester head, H73.

The latest family of harvester heads is shown in Figure 13.6. In addition, a special head has been developed to cut eucalyptus trees.

The products shown in the figures have their own configuration matrix to present the configuration knowledge related to the product families. The cranes are the easiest to configure, whereas harvester heads are very complex due to the large number of interdependent features that need to be selected during the configuration process.

Harvesters and harwarders are decomposed into base machine, crane and harvester head, whereas forwarders are decomposed into base machine and crane. All three sub-systems for the harvester are configurable; that is, they are modular platform architectures. All sub-systems have their own configuration models, and are configured separately and produced with their

FIGURE 13.2
Ponsse forestry machines. (From Nummela, J., Integrated Configuration Knowledge Management by Configuration Matrices—A Framework for Representing Configuration Knowledge to Organization, PhD thesis, Institute of Production Engineering, Tampere University of Technology, 2004. With permission.)

FIGURE 13.3
Ergo 8w. (Photograph courtesy of Ponnse. Reprinted with permission.)

FIGURE 13.4
Ergo and Elk in action. (Photograph courtesy of Ponnse. Reprinted with permission.)

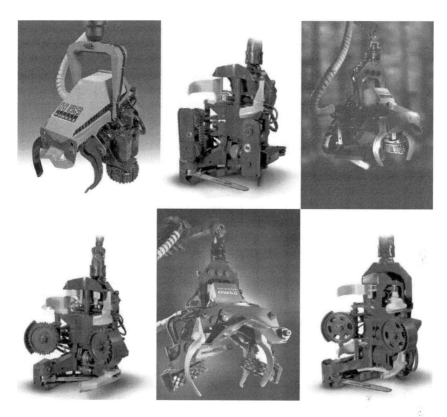

FIGURE 13.5
The family of harvester heads circa 2004. (From Nummela, J., Integrated Configuration Knowledge Management by Configuration Matrices—A Framework for Representing Configuration Knowledge to Organization, PhD thesis, Institute of Production Engineering, Tampere University of Technology, 2004. With permission.)

FIGURE 13.6
Ponsse's latest harvester heads H8 and H6 from the new harvester head range. (Photograph courtesy of Ponnse. Reprinted with permission.)

own shop orders. The product structures of all the configurable modules are decomposed into a functional structure as presented in Figure 13.7. The same concept is used in all products.

The products are configurable through functional modules. The functional modules are used for the option structures. However, the product platform is based on structural modularity. Product development designs the needed variants for product structures, but the decomposition of the entire product is mainly done in cooperation with the designer and the personnel from documentation. The output from the documentation department is part drawings. From these drawings production forms manufacturing modules using the same interfaces as depicted in the part drawings. A considerable amount of iteration between production and documentation occurs if there are problems with the module interfaces, which is particularly noticeable during prototyping and ramp-up. The conclusion is that even if the processes for Ponsse's product development are effective and flexible, the current configuration process generates modular structures that have ripple effects throughout the company. Manufacturing cannot work with functional structures, and final assembly cannot build functions because they work with structural modules. There is a mismatch between the product functional modules and manufacturing structural modules.

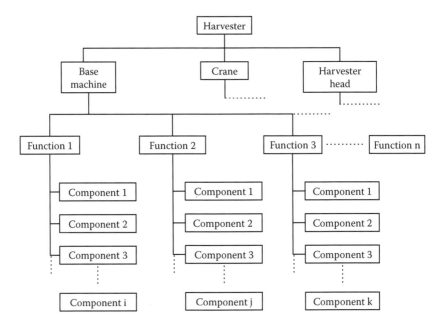

FIGURE 13.7
Functional modular product structure. (From Nummela, J., Integrated Configuration Knowledge Management by Configuration Matrices—A Framework for Representing Configuration Knowledge to Organization, PhD thesis, Institute of Production Engineering, Tampere University of Technology, 2004. With permission.)

Configuration Matrices Solution

The solution was to develop an automatic configurator based on configuration matrices. The matrices convert the functional product structure into a modular structure suitable for manufacturing.

The configuration knowledge was captured using charts. An example chart is shown in Table 13.5. The configuration matrix derived from the example chart is shown in Table 13.6.

TABLE 13.5

Example Chart for Capturing Configuration Knowledge

Base Machine—Variant Part of the Structure	
M00001	Hydraulics
M00002	Tools package
M00003	Clutch
M00004	Engine
M00005	Cables
M00006	Hydraulic pump

Options and Their Module Selections
If rotator XXX selected include:
| M00007 | Rotator |

If rotator XXY selected include:
| M00008 | Rotator |

If crane XXX selected include:
| M00009 | Crane |

If crane XXY selected include:
| M00010 | Crane |

If rotator XXX and crane XXX selected include:
| M00011 | Crane hydraulics |

If rotator XXY and crane XXY selected include:
| M00012 | Crane hydraulics |

If rotator XXY and crane XXX selected include:
| M00013 | Crane hydraulics |

If lights selected include:
| M00014 | Lights |
| M00015 | Lights |

If lights and rotator XXX and crane XXY selected include:
| M00016 | Lights |

Source: Nummela, J., Integrated Configuration Knowledge Management by Configuration Matrices—A Framework for Representing Configuration Knowledge to Organization, PhD thesis, Institute of Production Engineering, Tampere University of Technology, 2004. With permission.

Note: If rotator XXX selected, then crane XXY cannot be selected.

TABLE 13.6

Configuration Matrix Derived from Table 13.5

	Rotator XXX	Rotator XXY	M00007 Rotator	M00008 Rotator	Crane XXX	Crane XXY	M00009 Crane	M00010 Crane	M00011 Crane Hydraulics	M00012 Crane Hydraulics	M00013 Crane Hydraulics	Lights	No Lights	M00014 Lights	M00015 Lights	M00016 Lights
Rotator XXX			×				×	×	×		×			×	×	×
Rotator XXY				×			×	×		×				×	×	
M00007 rotator	×															
M00008 rotator		×														
Crane XXX							×		×		×			×	×	
Crane XXY								×		×				×	×	×
M00009 crane	×	×			×											
M00010 crane	×	×				×										
M00011 crane hydraulics	×				×											
M00012 crane hydraulics		×				×										
M00013 crane hydraulics	×				×											
Lights													×	×	×	×
No lights												×	×			
M00014 lights	×	×			×	×						×				
M00015 lights	×	×			×	×						×				
M00016 lights	×					×						×				

Source: Nummela, J., Integrated Configuration Knowledge Management by Configuration Matrices—A Framework for Representing Configuration Knowledge to Organization, PhD thesis, Institute of Production Engineering, Tampere University of Technology, 2004. With permission.

A total of 22 matrices were developed to cover the entire spectrum of products. The total number of modules was around 2,000, of which approximately 1,000 were variants (different from each other). The variants covered over 18,000 parts and components. There were 545 features with over 1,500 options in the 22 configuration matrices. The enterprise resource planning (ERP) system handles the conversion from functional to structural modules using routing functions.

Results

The results of using configuration matrices are shown in Table 13.7.

TABLE 13.7

Comparison of Before and After Configuration Matrices

Activity	Old Situation (Before)	New Situation (After)
Configuration task in sales	Manual (errors likely to happen)	Web-based configurator with valid configuration knowledge
Configuration task in manufacturing	Manual (errors likely to happen)	Automatic with valid configuration knowledge
Information from orders	E-mail sent manually (includes the contract done with the customer)	Automatic e-mails with a valid configuration specification (no work related)
Configuration models	Ill defined	Well defined
Configuration knowledge maintenance	Ill defined	Well defined
Configurator database update	Not applicable	Automatic with revision knowledge
Double-checking the configurations	Critically needed	Not needed
Revision changes	Not updated	Automatically updated
Version changes	Updated manually	Updated automatically with well-defined algorithms and processes
Change interval from product design	Continuous	Well-defined intervals (critical changes only established immediately)
Order book configurations	Not updated	Updated automatically with well-defined algorithms and processes
Customer changes	Reacted to with complex processes	Reacted to with well-defined algorithms and processes
Information from changes	Manually done in many steps	E-mails automatically to right people with the right information
Reconfiguration	Manual with tacit knowledge	Easily done with a configurator and revisioned databases

(Continued)

TABLE 13.7 (*Continued*)

Comparison of Before and After Configuration Matrices

Activity	Old Situation (Before)	New Situation (After)
Pricing	Basis not well established	Basis automatically established
Learning configuration models	Months	Minutes (only basis for configuration matrices needed)
Knowledge dependency on personnel	Heavily dependent	Not dependent at all
Wasted time related to configuration process	Over 90%	0%

Source: Nummela, J., Integrated Configuration Knowledge Management by Configuration Matrices—A Framework for Representing Configuration Knowledge to Organization, PhD thesis, Institute of Production Engineering, Tampere University of Technology, 2004. With permission.

Exercises and Problems

1. Describe a configuration model.
2. Describe a configuration matrix.
3. Explain the relationship between a configurator and a configuration matrix.

Reference

Nummela, J. 2004. Integrated configuration knowledge management by configuration matrices—A framework for representing configuration knowledge to organization. PhD thesis, Institute of Production Engineering, Tampere University of Technology.

Section IV

Synthesis—People

14

The Socio-Technical Systems Approach

Introduction

The foundations for studying and designing industrial organisations as socio-technical systems were laid by social scientists at the Tavistock Institute of Human Relations, London, England. Most of the conceptual framework was developed in the period 1950–1970. Unfortunately, this work seems to have been forgotten today. Yet it is important because it provides a scientifically proven basis for designing work groups that are productive and jobs that are meaningful and satisfying.

The socio-technical systems approach developed over a number years through a series of action research field studies. A brief overview of these studies is given in Cummings and Srivastva (1977, Chapter 4), and a detailed overview and discussion in Emery (1978). The concepts and procedures are general and applicable to all industries employing technology.

Cummings and Srivastva (1977) provide an in-depth explanation of the approach with two detailed examples. Taylor and Felten (1993, Chapter 2) give an overview of the socio-technical systems approach in the United States.

Figure 14.1 shows the concept map for this chapter. The chapter explains the key concepts of socio-technical systems, and presents design guidelines and a process for designing, implementing and operating them. The chapter is linked with group technology in Chapter 4. Group technology layouts have task relationships that facilitate better work relationship structures; thus they complement socio-technical design principles. The organisational structure of a small to medium manufacturing company (SMMC) should be designed to support socio-technical work groups and integrate them effectively with other work groups. This aspect is discussed in Chapter 15.

The concept *socio-technical system* encompasses the key ideas described below (Emery 1978).

Substantive Components

The elements of the system are substantive (real). They are people, machines, equipment, etc. A socio-technical system will have political and economic

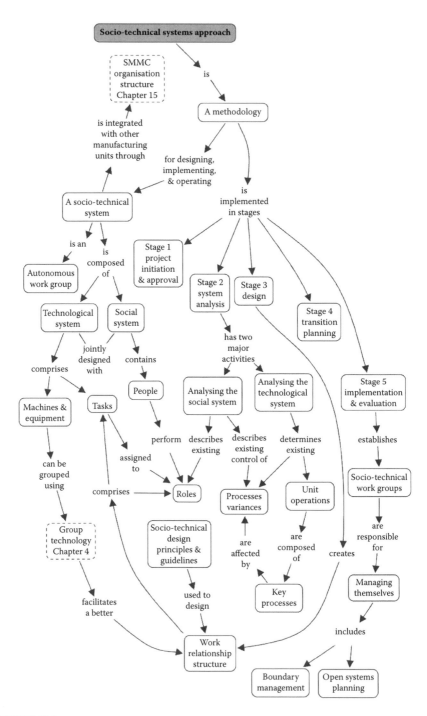

FIGURE 14.1
Concept map for Chapter 14.

aspects that can be studied as systems, but they are abstract systems. That is, a political or economic system is formed by abstracting certain features from real social systems. The difference is critical. Social systems are real: their elements are people. Social systems encompass all aspects of human behaviour. On the other hand, political and economic systems are not real: their elements are abstract features of social systems. Social scientists made this distinction clear by taking a person as the object of analysis and defining the socio-technical system as the social unit of analysis.

Open System

A company is studied as an open system. Prior to the concept of socio-technical systems, nearly all social scientists studied social systems using closed system models. There is a big difference between closed and open system models (Chapter 2). Open system models focus on the environment and system-environment relationships as these affect what the system can and cannot do. Open systems can also evolve and adapt over time. These aspects are not included in closed system models.

Two Subsystems: Technological and Social

A company is viewed as consisting of two sub-systems: a technological system and a social system—hence the name. A social system is a system comprised solely of people. The technological system consists of tools and machines.

> It might be practically justifiable to exclude the technological component from the system concept if it were true, as many writers imply, that it plays a passive and intermittent role. However, it cannot be dismissed simply as a set of limits that exert influence only at the initial stage of building an enterprise and at such subsequent times as an enterprise oversteps these limits. There is an almost constant accommodation of stresses arising from changes in the external environment, and the technological component not only sets limits upon what can be done, but also creates demands that must be reflected in the internal organization and ends of an enterprise. (Emery 1978, p. 42) (*Note*: The term enterprise means the same as industrial organisation.)

The concept *socio-technical system* serves as a frame of reference. It directs attention to the technological system and the social system, their interrelationship, the external environment and the relationship between the environment and the joint socio-technical system.

The two sub-systems are linked together through work roles. Tasks required in the technological system are combined into roles. The roles are linked to each other and create the work relationships in the social system. There is no one-to-one relationship between the tasks performed in the technological system and the structure of the social system. It depends on how the roles are designed.

Work Roles

A role is a set of expectations about how a person is to behave in a social group or company. It is a simplified description of behaviour, like a part in a play. A role might be performed by different people at different times, but provided they meet the role requirements, the role is the same. However, the effectiveness of role performance depends on the person acting in the role. A role cannot be defined without taking into account other roles. Roles have to be designed as a system.

In home life, the roles of mother, father, etc., are developed through intense, interpersonal relationships over a relatively long period of time. On the other hand, explicit (codified) role definition in a company enables people to coordinate and integrate their behaviour without having to be in continuous contact and without them having to know each other personally. These work roles enable companies to operate effectively with people separated geographically and in time (shift work). In practice, work roles are a mixture of prescription (explicit role definition) and interpersonal relationships. The interpersonal relationship and communication interactions increase as roles move from more to less explicit prescription. So an advantage of explicit role definition is a reduction in the degree of communication.

Roles are dynamic. They need to be maintained and developed. Role maintenance requires the person in the role to act in a way that reaffirms the expectations associated with the role. For example, consider a male doctor with a female patient. During examination the doctor, by his actions, must assure the female patient that he is looking at her as a doctor and not as a potential lover. This affirmation is required each time the patient visits.

Roles can and should change over time. This is role development. The person in the role will modify his or her behaviour to either change other people's expectations or to meet changes in other people's expectations. The changes may be small adjustments or abrupt, large changes. A minor change might be someone giving up smoking. An abrupt change would be a criminal turning to religion and leading a law-abiding life.

What is work? Work is an agreement between two or more people, in which one party agrees to perform stated tasks for the other party (Cummings and Srivastva 1977). The parties may be individual people, social groups or companies. Work may be performed free of charge or for a fee. An employee is a person who performs work for a company for a fee.

A work role is the set of expectations associated with the agreement to perform tasks. As there are two parties to the agreement, there are two sets of expectations associated with a work role (Newman 1973). Both viewpoints are valid and need to be taken into account when work roles are designed (Figure 14.2).

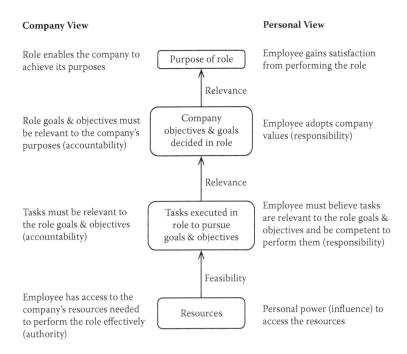

Company View   Personal View

Role enables the company to achieve its purposes — **Purpose of role** — Employee gains satisfaction from performing the role

Relevance

Role goals & objectives must be relevant to the company's purposes (accountability) — **Company objectives & goals decided in role** — Employee adopts company values (responsibility)

Relevance

Tasks must be relevant to the role goals & objectives (accountability) — **Tasks executed in role to pursue goals & objectives** — Employee must believe tasks are relevant to the role goals & objectives and be competent to perform them (responsibility)

Feasibility

Employee has access to the company's resources needed to perform the role effectively (authority) — **Resources** — Personal power (influence) to access the resources

FIGURE 14.2
Definition of a work role.

The company's expectations are as follows:

- The purpose of the work role is to enable the company to achieve its purposes.
- The objectives and goals of the role should be relevant to the company's purposes and in accordance with its set of values. The company holds the employee *accountable* for achieving the goals and objectives to the required standards. That is, the employee will be rewarded if the goals and objectives are achieved and punished otherwise. Accountability is necessary to ensure the employee performs as expected.
- The tasks should be relevant to the pursuit of the goals and objectives of the role and in accordance with the company's set of values. The employee is expected to perform the stated tasks to the satisfaction of the company. It holds the employee accountable for the role tasks.
- Company resources are made available to the employee so that the tasks can be successfully performed. The company grants the employee authority: the right of access to or control over its resources. Provision of company resources is necessary for the tasks to be feasible. An employee cannot be held accountable for role performance if the resources required for the role are not provided.

The employee's expectations are as follows:

- The work role is one of many different roles the employee has in life. For example, other roles of a male employee might be husband, father, son and member of a professional association. The employee's purpose in acting in a work role is to gain satisfaction, directly or indirectly. Direct satisfaction comes from performing the tasks in the work role. Indirect satisfaction comes from the rewards for performing the work role (e.g., money), which can be used to perform other roles in life. The employee has to integrate the purposes of the work role with the purposes of the other roles in life.

- The employee has to effectively pursue the goals and objectives of the work role. At the same time, the employee will be pursuing other goals and objectives for other roles. The employee has to integrate the work role goals and objectives with those for other roles. The degree of integration will affect the willingness and persistence of the employee to pursue the goals and objectives. This is the responsibility the employee has for the role. An employee is responsible if he or she rewards himself or herself for successful performance of the tasks and punishes himself or herself otherwise. It is very important to be clear about the difference between accountability and responsibility. Accountability is defined in terms of the power of the company to punish the employee. On the other hand, responsibility is personal; it depends on the person filling the role. A company cannot force an employee to be responsible. It can hold an employee accountable for a role, but he or she may not feel responsible. Good role design encourages responsibility; bad role design does not.

- The employee has to adopt the company's values in terms of choosing how to pursue the work role goals and objectives. He or she has to integrate these values with those for other roles in his or her life. The degree of integration will affect the willingness and persistence of the employee to carry out tasks. This is the responsibility the employee has for the role.

- The employee will have a degree of willingness and ability to use company resources that have been made available. This is the power of the employee to fulfil the role.

From the company's point of view, the employee in a work role is expected to perform the tasks necessary to achieve the work role goals and objectives and use its resources efficiently and effectively. The company requirements are constraints on the employee. They limit what the employee can do. On the other hand, they also enable an employee to achieve more than what he or she could on his or her own. In this sense the requirements increase the power of the employee.

An individual work role does not exist in isolation. All the work roles in a company are interconnected. Thus each work role is constrained by others that affect it and those that are affected by it. Roles have to be designed as a system. In addition, there may be constraints on roles arising from the environment, e.g., government laws and regulations.

Work roles may be performed ineffectively for the following reasons:

- The role has not been properly defined or the employee has not been properly informed of the role requirements.
- Company goals and objectives conflict with the employee's goals and objectives.
- The role goals and objectives conflict with those of other roles in the company, or outside constraints, and the employee has no means to resolve these conflicts.
- The employee does not believe the stated tasks will achieve the company's goals, and hence feels little responsibility for effective task performance, or the nature of the tasks and the rewards may produce a feeling of lack of responsibility in the employee.
- The employee is not competent to perform the tasks.
- Company resources made available to the employee are inappropriate or inadequate to achieve the tasks. That is, the authority does not match the accountability.
- The employee does not have the power to use the resources.

The design of jobs (work roles) and the structuring of these to form work groups creates a work relationship structure, a set of work roles and their interrelationships (Figure 14.3). Once in place, this structure constrains the behaviour of both the employees and the supervisors or managers in charge of them. A good work relationship structure permits effective task performance. A bad one does not, and leads to adverse effects, such as high rates of absenteeism and employee turnover, frequent accidents, high product reject rates, etc. What is a good work relationship structure? How do we go about designing such a structure? The socio-technical systems approach answers both questions.

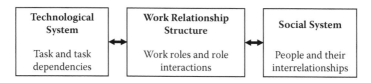

FIGURE 14.3
Concept of work relationship structure.

Analysing the Technological System

The first step in analysing a system is determining what it is supposed to do. Companies have two basic objectives (Newman 1973). One is what the company is set up to do (output objective), and the other is to achieve what enables it to continue (capacity objective). Over time both objectives will need equal amounts of attention. But at any one time one objective may be receiving more attention than the other. So the first requirement is to determine the output objective and the main process contributing to it. The capacity objective, though essential, is not the reason for the existence of the company: profit is not an output objective.

The process is described as an input-conversion-output cycle. It is called the primary task. The inputs for SMMCs are materials that are converted to processed materials, components, sub-assemblies, assemblies, products and systems. The material conversion process is dominant for discrete part manufacturing.

The purpose of a system and its primary task depend on the purposes of the person analysing the system. The traditional socio-technical system approach assumed that the people conducting the analysis are the employees and management or consultants working on their behalf. Hence they determine the primary task. Current approaches include all stakeholders.

The grouping of people in the social system depends on the phases of production of the primary task. Step 2 determines what these phases are. The phases are referred to as unit operations (Figure 14.4). Process flowcharts are used to describe the primary task and identify the unit operations.

For example, unit operations for a group technology layout might be:

- Op 1: Presswork
- Op 2: Machining

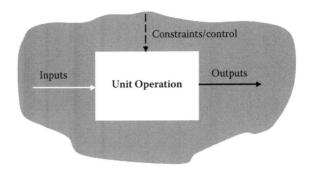

FIGURE 14.4
Concept of unit operation.

- Op 3: Finishing (heat treatment, plating and riveting)
- Op 4: Assembly

An example for brickworks is:

- Op 1: Premixing
- Op 2: Extrusion
- Op 3: Firing

Unit operations are connected together as illustrated in Figure 14.5.

The aim of step 3 is to identify the key process variances and their interrelationships. These are defined in a matrix of variances (Table 14.1). The variances form the columns and rows of the matrix. They are ordered according to the unit operations shown in the right-hand column. The most important (key) variances are indicated with the letter K on the diagonal; otherwise an X is entered. Interrelationships between variances are indicated by an X below the diagonal, which means the column variance affects the row variance.

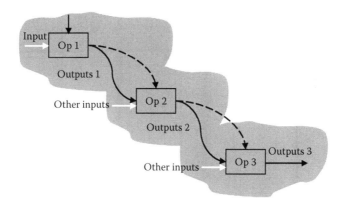

FIGURE 14.5
Illustration of three connected unit operations.

TABLE 14.1

Matrix of Variances

Process Variance	a	b	c	d	e	Unit Operations
a	X					Op 1
b		K				
c	X		X			Op 2
d		X		K		
e	X				X	

Analysing the Social System

The purpose of social system analysis is to describe the main characteristics of the existing social system. It begins with a review of the organisational chart. Next, the chart is described in more detail in terms of the number of organisational levels, types of groups and current work roles.

The next step is to develop a table of variance control (Table 14.2). The purpose of the table is to answer the following questions (Emery 1978, p. 98):

- Where does variance occur in the process?
- Where is it observed?
- Where is it controlled?
- Who controls the variance?
- What tasks have to be performed to control it?
- What information is needed and from where can it be obtained in time to carry out the control activities?

There is an additional column for hypotheses for redesign, as it has been found worthwhile to write these as they arise; otherwise they can be easily forgotten later on.

Other activities undertaken during this stage include:

- Describing the ancillary activities (other activities required for successful completion of those associated with controlling the key variables)
- Mapping out the physical or geographic and timing relationships between roles
- Determining the flexibility of work roles
- Defining the payment system
- Determining workers' perceptions of their roles
- Evaluating roles against a list of psychological needs

Designing the Work Relationship Structure

Characteristics of the technological system that have a major impact on the social system and need to be taken into account when designing the work relationship structure are:

- The characteristics of the material being produced
- The physical work setting

TABLE 14.2

Table of Control of Key Variables

Key Variable	Name of Unit Operation		Controlled by Whom	Control Tasks	Information Needed for Control	Information Source	Hypotheses for Re-Design
	Occurs Where	Observed Where	Controlled (yes/no) and Where				

- The spatial layout and the spread of the production process over time (effect of shifts)
- The level of automation
- The manufacturing operations and the grouping of these operations into phases of production
- The degree of importance of different operations

These characteristics affect the kinds of tasks that need to be performed, the skills and abilities required to perform the tasks, and task dependencies. The latter is particularly important. Independent tasks do not require cooperation between employees. Hence tasks can be grouped to minimise needs for external control (similar work roles assigned to a supervisor), to maximise the delivery of support services (e.g., maintenance), or to reduce the distance over which information or materials must travel.

Dependent tasks require cooperation between employees. A major constraint on the coordination of dependent tasks is the degree of interdependence between tasks. Task dependencies can be of two forms: simultaneous and successive. Simultaneous interdependence arises when a task is broken into part-tasks that must be carried out jointly. The employees performing the part-tasks must coordinate their activities simultaneously, e.g., two workers lifting a carton of parts.

Successive interdependence is affected by the time lag between tasks and the fact that the tasks may not be of the same size or duration. The time lag between tasks can be increased by using buffer inventories. The task sequence in one job must be of the same duration as the task sequence in the preceding and succeeding jobs because the cycle time is the same. If the durations are different, then workers with short-duration jobs have to wait for the worker with the longest-duration job to complete his or her tasks, lowering productivity. Work flow balancing arranges tasks to ensure they are approximately the same size and duration.

The following design guidelines for combining tasks into roles (jobs) at the individual and group levels are based on Emery (1978), Davis (1957) and Rice (1958).

Individual Level

- Optimal task variety. Tasks should not have too much or too little variety. Too little variety leads to boredom or fatigue. Too much can lead to frustration and reduces the efficiency of training. Jobs should provide some "stretch" tasks mixed with routine tasks. The appropriate mix varies from individual to individual and between cultures.
- Optimal cycle time. The cycle time should not be too short or too long. A cycle that is too short reduces productivity because a large

portion of the cycle is taken up in starting and finishing. A cycle that is too long makes it difficult to build a rhythm of work. Both cases reduce the learning that can be achieved.

- Interdependent tasks. The tasks in a job should be interdependent, such that completion of one task improves the quality or productivity of the successive or co-dependent tasks. That is, the tasks should form a system that is a more complex task. This allows a worker to find a method of working that best suits his or her requirements. If all tasks are designed this way, then workers can more easily relate their jobs to each other, improving group productivity.

- Worker involvement in setting standards and feedback of results. "Minimum standards generally have to be set by management to determine whether a worker is sufficiently trained, skilled or careful to hold the job. Workers are more likely to accept responsibility for higher standards if they have some freedom in setting them and more likely to learn from the mob if there is feedback. They can neither effectively set standards nor learn if there is not a quick enough feedback of knowledge of results" (Emery 1978, p. 87).

- Job enlargement. Include auxiliary, preparatory and control tasks in the job. These tasks have an important effect on the main job tasks. "The worker cannot and will not accept responsibility for matters outside his control" (Emery 1978, p. 87).

- Community value. The tasks included in the job should require knowledge and skills that the community respects.

- Perceivable product contribution. The job should make a contribution to the product that can be clearly seen.

- Meaningful incentives. Provide incentives that are important to the worker.

Group Level

- Interdependent jobs. Where jobs are interdependent, provide "interlocking" tasks between jobs, job rotation or place the jobs close to each other. "At a minimum this helps to sustain communication and to create mutual understanding between workers whose tasks are interdependent, and thus lessens friction, recriminations, and 'scapegoating'. At best this procedure will help to create work groups that enforce standards of cooperation and mutual help" (Emery 1978, p. 87).

- High-stress jobs. If the individual jobs are likely to produce high levels of stress, then provide interlocking tasks between jobs, job rotation or place the jobs close to each other. This allows workers to relieve stress through mutual support.

- No perceivable product contribution. If the individual jobs do not make a perceivable contribution to the product, then provide interlocking tasks between jobs, job rotation or place the jobs close to each other.
- Linked jobs. If jobs are linked together by interlocking tasks or job rotation, according to the three conditions above, they should as a group have:

  (i)   a semblance of an overall task which makes a contribution to the utility of the product;
  (ii) a some scope for setting standards and receiving knowledge of result;
  (iii) a some control over the "boundary tasks" (Emery 1978, p. 88)

- Group stability. Ensure the range of skills within a group is such that all workers in the group can understand and acquire them. Communication becomes more difficult with greater skill differences and reduces group cohesiveness.
- Work group size. Rice (1958) argued that the optimal group size was 6–12. But as Cummings and Srivastva (1977) correctly point out his argument is based on clinical experience of "unstructured" groups. Structuring tasks into work roles (jobs) reduces the communication requirements for effective performance. Other studies have shown that 40 member work groups can be effective. Thus size is not an important criterion in itself. The correct size is one that groups tasks to minimise the number of workers without conflicting with other design guidelines.
- Communication channels. Provide communication channels to allow workers to input their requirements at an early stage in the design of new jobs.
- Foreman/supervisor sanctioned promotion. Provide means for workers to approve (sanction) promotions of their foremen and supervisors.
- Work group misfits. Provide a means to allow a worker to move to another work group if he or she does not fit into the current group.

Emery (1978, p. 88) states that "these requirements are not confined to operators on the factory floor, nor is it possible to meet them in the same way in all work settings or for all kinds of people."

The guidelines above provide an initial group structure that can learn and adapt. The groups are autonomous in the sense that all the important work variables are under the group's control, including interactions with other work groups. The groups do not need a supervisor. They manage themselves. Furthermore, they have the capability to learn and evolve over time without management intervention.

Managing the System–Environment Relationship

The environment of a work group consists of other work groups in the company, management, and may also include people and organisations outside the company, e.g., suppliers and customers. Each work group is responsible for managing its environmental relationships. There are two major activities involved in managing these relationships: boundary management and open system planning (Cummings and Srivastva 1977).

Boundary Management

Boundary management consists of two activities: protecting the primary task and regulating the environmental exchanges (Cummings and Srivastva 1977). The boundary of the work group is determined by the boundary of the primary task, which can be differentiated by technology, territory, time or a combination of these. Protection gives the work group a certain degree of independence from its environment. It can be protected using any or all of the following strategies (Thompson 1967):

- Buffering: Set up input and output boundary elements. Buffering of materials and components is achieved through input and output inventories. Buffering prevents learning because it absorbs environmental variation. The variation is hidden in the buffer. The optimal buffer size is a value that absorbs some but not all variation. It absorbs enough to ensure smooth performance of tasks within the work cycle, with little interruption. It allows some variation so the work group can learn and adapt to improve performance.

- Levelling: Smooth the input and output exchanges. Smoothing acts on the environment to reduce variation before it can reach the work group boundary.

- Anticipating: Forecast and adapt to environmental changes. Environmental variation is detected early and plans are made to counter it before it reaches the work group boundary.

- Rationing: If the three methods above fail, then rationing is used. In this situation, environmental variation has overloaded the system and it cannot effectively deal with it within the standard work cycle.

Boundary exchanges penetrate the environment and the system, as discussed in Chapter 2. There will be at least two people involved in the exchange: one in the environment and one in the work group. They should be jointly responsible for the exchange. Together they should work out the best method for handling the exchange.

Open Systems Planning

Open systems planning is a development function of the work group. It aims at finding and interpreting the patterns occurring in the environment (L_{22} processes) and the impacts these are likely to have on the work group in the future. The work group can then decide to restructure to meet these changes or plan to influence the environment to change in directions deemed desirable to the group. Open systems planning can be carried out in four steps (Cummings and Srivastva 1977):

1. Identify and describe the dynamic processes (L_{22} processes) in the environment. These will include social, technological and ecological processes. Only include those processes that have the most effect on the primary task.
2. Identify and describe how the work group currently acts toward and values these processes.
3. Determine how the work group wants to respond to these processes in the future.
4. Make plans to restructure the group to adapt to the processes or to influence them to change in desired directions.

The socio-technical guidelines produce work groups that are autonomous. Management interventions in the work groups should be restricted to setting work goals and standards and that is all. Work groups should be allowed and given the resources to manage themselves and the work group–environment relationships. The advantages gained from using autonomous work groups are: (1) job satisfaction is higher, (2) the work groups perform more effectively, and (3) supervision costs are reduced.

Implementing Socio-Technical Systems

Changing from a traditional organisation to autonomous work groups is a major undertaking. The change process must be supported by senior management and carefully planned. A generic process for implementing socio-technical systems is shown in Figure 14.6.

Project Initiation and Approval

The project begins with a discovery stage where the company becomes aware of the principles and practices of socio-technical systems. Awareness is necessary, but insufficient. There has to be a champion. The champion

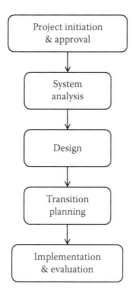

FIGURE 14.6
Process for implementing socio-technical systems.

is a person in the company who is prepared to sell the socio-technical system concept to management and get their approval. Champions can come from anywhere in the company. However, the successful champion will be a person who has the confidence of senior management. After the champion has been identified, the next step is to gain senior management approval for the project.

All change meets with some resistance from the people involved. Resistance can be significantly reduced by starting with a pilot project and disseminating the successful results to prove the effectiveness of the change. So the next step is to set up a pilot project. This involves defining the scope of the project, setting up the initial project organisation (action group) and preparing a preliminary plan of work.

The scope is determined by selecting the unit in the company for the pilot study. This could be a complete factory or a department in a factory. The criteria for selecting the pilot unit are (Cummings and Srivastva 1977):

1. The boundary can be clearly defined.
2. Its inputs and outputs can be clearly defined and measured.
3. There is a high probability of success. Ideally, the unit should have low or mediocre performance or morale. The chances of success are higher in this case, as the members of the unit have little to lose and much to gain.

4. High potential for disseminating results. The unit should have characteristics that can be generalised to the other parts of the company.

5. The members of the unit are interested to try out the socio-technical system approach.

Possible units should be identified and compared according to the criteria. It is unlikely that one unit will score highest on all criteria. When no unit scores highest on all criteria, the final decision can be based on criteria 3 and 4, as these are the most important.

After the unit is selected, the next step is to sanction the project. First, the members of the sanctioning body are selected. Membership should include the key stakeholders, e.g., managers of the unit, managers of related units and worker representatives. The purpose of the sanctioning body is to define the sanctioning rules that will protect the unit during the pilot project. Rules need to be determined for three types of protection: conceptual, experimental and operational. Conceptual protection allows workers to freely explore all aspects of their work and work environment during the re-design, promoting innovative solutions. Experimental protection is needed to protect workers during the pilot project. They cannot be expected to meet full operational load during the trials. They must be given time to learn the new system. Operational protection is required during ramp-up. Again workers must be given time to learn the new system and diagnose and resolve problems that occur during ramp-up.

The next step is to form an action group. It is responsible for supervising the analysis, re-design, implementation and evaluation of the project and the

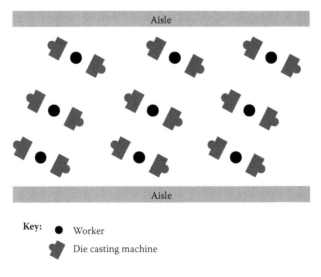

FIGURE 14.7
Layout of a zinc die-casting department.

dissemination of the results. The action group will consist of key managers and workers of the unit. It may also include outside consultants. For effectiveness the group should be small, with around five or six members. It reports to the sanctioning body.

The final activity in this phase is to prepare a preliminary plan of activities for completing the pilot project. The deliverables for this phase are:

- A socio-technical systems champion
- Senior management approval of the project
- Pilot project specification, organisation and plan

System Analysis

The analysis phase includes technological and social system analyses, and an open system scan that identifies boundary problems and describes the key characteristics of the environment.

The deliverable of this phase is an analysis report containing the following:

- Objective of the system
- Description of the primary task and its unit operations
- Matrix of variances
- Table of variance control
- Organisation chart
- Description of ancillary activities
- Map of role relationships
- Description of role flexibility
- Description of payment system
- Survey of worker's perceptions of their roles
- Role evaluation against psychological needs

Design

In this phase, the unit is re-designed using socio-technical system design guidelines. The deliverable for this phase is the new design.

Transition Planning

In this phase, a detailed action plan is created for changing from the current system to the new design. The plan should take into account transitional protection, training needs, changes in supervisory roles, changes in incentive schemes, and allow time for feedback and evaluation of the changes. The deliverable for this phase is the action plan.

Implementation and Evaluation

The final phase includes the implementation of the action plan, ramp-up to full production, and evaluation and dissemination of the results. The deliverables are:

- Fully functioning autonomous work group operating at full operational capacity
- Evaluation report
- Disseminated results

Exercises and Problems

1. Briefly define a work role.
2. List all the roles you perform in your life. Rank them in terms of importance to you. Now estimate the time you spend in each role and rank the roles according to time. Compare the two rankings.
3. Briefly define a work relationship structure.
4. List the steps to implement a socio-technical system.
5. a. How is the socio-technical system-environment relationship managed?

   b. Who is responsible for managing the relationship? Explain your answer.
6. Figure 14.7 shows the layout for a zinc die casting department. The department makes zinc alloy wheels for sports cars. The workers' job is to unload the die casting machines and to stop them if a problem occurs. The die casting machines are filled with molten zinc from a very large container using an overhead gantry. The pots holding the zinc are indicated by the semi-circle at the back of the machine. Each worker looks after two machines. The workers have to turn around to unload the machines, as can be seen from the figure. Evaluate the layout using socio-technical design principles. Develop a new design concept to improve the layout.

References

Cummings, T.G., and Srivastva, S. 1977. *Management of work: A socio-technical systems approach.* San Diego: University Associates.

Davis, L.E. 1957. Toward a theory of job design. *Journal of Industrial Engineering*, 8, 305–309.

Emery, F.E. 1978. *The emergence of a new paradigm of work*. Canberra: Centre for Continuing Education, Australian National University.

Newman, D. 1973. *Organization design*. London: Edward Arnold.

Rice, A.K. 1958. *Productivity and social organization: The Ahmedabad experiment*. London: Tavistock Publications.

Taylor, J.C., and Felten, D.F. 1993. *Performance by design: Sociotechnical systems in North America*. Englewood Cliffs, NJ: Prentice-Hall.

Thompson, J.D. 1967. *Organizations in action*. New York: McGraw-Hill.

15

Organising People for Manufacturing

Introduction

The focal concepts for this chapter are *social systems* and *interactive planning*. The chapter looks at how small to medium manufacturing companies (SMMCs) are designed as social systems (Figure 15.1). It shows the different ways work roles can be effectively linked together to form a superior-subordinate hierarchy (organisational structure) and talks about the different kinds of work groups and teams and how they can be linked to the organisational structure. It also describes a powerful planning process called interactive planning that will help SMMCs successfully adapt to turbulent environments.

Before we discuss how SMMCs should be designed we need to understand their purpose and function in a society. A SMMC is wealth producing (profit making). It generates and distributes wealth at the same time; these are its two main functions (Chapter 10). Thus SMMCs help members of a society progress toward the ideal state of plenty. Some people, including the authors, are questioning this narrow focus and argue that SMCCs should facilitate progress toward all four ideals. Section II briefly showed how SMMCs can do this. The change in focus from one to four ideals increases the opportunities for a SMMC, increases its robustness, and increases the number and variety of stakeholders that will support it.

A SMMC affects its employees and people and groups who do not belong to it. These people (the stakeholders) can affect the overall performance of the SMMC by withdrawing support or actively fighting against it. To prevent this from happening SMMCs can include stakeholders in their decision-making process. The stakeholders are usually identified by exchange relationships (Ackoff 1981). An exchange relationship is one in which goods, services or money is provided in order to obtain other goods, services or money. The major types of stakeholders for a SMMC are customers, suppliers, government, investors, debtors and employees.

The traditional stakeholder viewpoint is based on economics and ignores other ways in which people can relate to SMMCs. For example, environmentalists may instigate a protest against a company for polluting a river. In this case, their motive is not financial but to protect the environment.

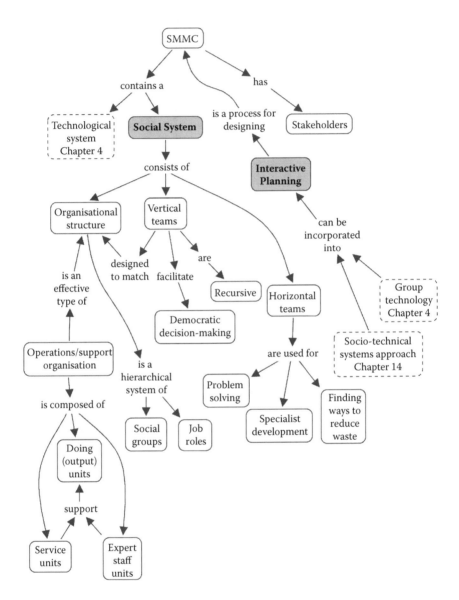

FIGURE 15.1
Concept map for Chapter 15.

The stakeholder viewpoint needs to be enlarged to include all people affected directly or indirectly by a SMMC.

Mitroff (1998) proposes a stakeholder viewpoint based on politics. He provides two classifications of stakeholders one based on stance and one based on function. Typical stance stakeholdlers are ally, collaborator, enemy and friend. Function role stakeholders include competitors, legal experts, researchers and regulatory agencies. The function role stakeholders also include people defined by exchange relationships.

SMMCs are usually designed to meet the needs of one or a few of the stakeholders, e.g., the shareholders. This is a dangerous practice, as other neglected stakeholders may become so aggrieved that they take action to disrupt operations. It is best to use the extended stakeholder viewpoint and serve as many different types of stakeholder as is feasible. Stakeholders can be included in the planning and redesign of the company, as discussed in the next section.

Interactive Planning

Interactive planning (Ackoff 1981, 1999) improves SMMC performance significantly because it aims at redesigning the SMMC. It can also be used to guide SMMCs and society toward an omnicompetence-based society (Britton 1989). Table 15.1 contrasts an omnicompetence-based society with Mumford's (1971) description of a power-based society. The former is based on power-to, the latter on power-over.

The basic principles of interactive planning practice are:

- Participative planning: The main benefit of planning is doing it. Process is more important than the product.
- Coordinated planning: All aspects should be planned simultaneously at any given organisational level.
- Integrated planning: Planning at different levels needs to be integrated. Normative, strategic, operational and tactical planning are interdependent.
- Continuous planning: Planning is a continuous process. Plans must be checked and updated frequently.

Interactive planning is a continuous process consisting of five phases: mess formulation, ends planning, means planning, resource planning, and implementation and control of the means plan. Each phase is itself a continuous activity and interacts with the other phases. Consequently, there is no fixed sequence for executing the phases, although most commonly it starts with mess formulation or idealised designing (ends planning).

Mess Formulation

The main purpose of mess formulation is to get management to move away from solving individual problems and issues as they occur and toward

TABLE 15.1

Comparison between Omnicompetence- and Power-Based Societies

Omnicompetence	Power
Key characteristics:	Key characteristics:
1. All have choice.	1. One or a few have choice.
2. Balanced pursuit of all four ideals: plenty, truth, good and beauty.	2. Pursuit of plenty and truth dominates pursuit of good and beauty.
Major Components	
5Cs	5Ps
Competence (power for all)	Power (for a few)
Choice	Profit
Consensus	Prestige
Cooperation	Productivity
Commonweal (communal property)	Property (private)

redesigning the SMMC. The reason for doing so is simple. Getting rid of something that you do not want does not necessarily give you what you do want. Though this may sound very reasonable, managers may be reluctant to completely redesign their company, especially if it is performing reasonably well. On the other hand, managers are more receptive to radical ideas if they are in a crisis and are not sure how to get out of it.

The first part of mess formulation involves a search for values in the social field and how these may affect the SMMC. The search for values can be conducted using a search conference (Emery and Devane 1999). New potential stakeholders will be identified during the search, and they should be invited to participate in the planning process.

Two kinds of analyses are performed during mess formulation: systems analysis and obstruction analysis. The systems analysis provides a thorough and concise description of the SMMC. The techniques described in Chapter 2 can be used for this purpose.

Obstruction analysis is performed to identify behaviour that prevents a SMMC from developing effectively. There are two kinds of obstructions: discrepancies and conflict. A discrepancy is a difference in belief between stakeholders about the SMMC and its environment. Discrepancies prevent development because the stakeholders have different views of the situation, and hence value differently the cost and benefits of development proposals. Discrepancy analysis lists the major differences in beliefs.

Some conflict within a SMMC is unavoidable. However, repetitive conflicts that cannot be resolved hinder development and create psychological stress. Conflict analysis identifies the major repetitive conflicts within a SMMC and their impacts.

The analyses are presented in a reference projection. This is a hypothetical story projecting how the SMMC will perform in the future assuming it does not change its mode of operation and the environment remains the same. The projection is not a forecast of the future. The projection should clearly demonstrate to management that they need to change the mode of operation in order for the SMMC to survive or thrive in the future.

Summing up, the aims of mess formulation are to:

- "Demonstrate to the members of the organization that the problems facing the organization are largely of their own making, and that significant improvement in organizational performance requires a major change in the way the organization operates." (Britton 1989, p. 252)
- Search for changes in values in the social environment and identify potential stakeholders.
- Convince management to involve all stakeholders in the interactive planning process.
- "Obtain consensus amongst the stakeholders with respect to the mess formulation. If the stakeholders agree with the mess formulation, then commitment to carrying out an idealized design is more easily obtained" (Britton 1989, p. 252).

The key feature of this phase is *consensus*.

Ends Planning

The purpose of ends planning is to specify long-term development targets (ends) of the SMMC. First, an idealised design is created (Ackoff 1981, 1999, Ackoff et al. 2006). Idealised designing is the most powerful method we know for making radical changes in a SMMC because it creates a vision of a better future. An idealised design is a design of the SMMC that the stakeholders would like to have. The design should encourage and enable all stakeholders to seek ideals (Britton 1989).

The design must comply with two constraints and two requirements. The constraints are:

- It must be technologically feasible. The technology on which it is based must currently exist or be available within the next year or so. This separates designs from science fiction fantasies and daydreaming. However, a design can be politically and economically infeasible. The reason for this is that political and economic feasibility depend primarily on people's values: change these and you change the feasibility. The purpose of idealised designing is to free people from existing constraints so that development can be more radical. "Other phases of interactive planning deal with how to generate political and economic commitment for the proposed design and with how closely it can be approximated." (Britton 1989, p. 252)

- It must be capable of surviving if brought into existence. The design must meet regulatory requirements and have the functions to operate and survive as a viable business. In addition, the sub-systems of the unit being designed have to be properly integrated. Integration is more important than optimising the performance of each sub-system.

The first requirement is that it must be able to adapt and learn. This is necessary for two reasons. First, the design is based on current conceptions of what is possible, but conceptions change over time. Second, all designs are necessarily incomplete. We do not have full knowledge about how to design companies. The design must allow for learning so a company can evolve over time.

The second requirement is the design should include a statement of purpose (Chapter 5).

Designs can be constrained or unconstrained. The unit being designed will normally be part of a larger unit. For example, a factory is part of a manufacturing company and a foundry is part of a factory. A constrained design assumes the larger unit is fixed and cannot be changed. In an unconstrained design, it is assumed that the larger unit can be changed. The unconstrained design will include design changes for the larger unit that will benefit the unit being designed. These design changes have to be proposed to the stakeholders of the larger unit for approval. If approved they become part of the design. All designs should be approved by consensus.

Idealised design teams should not exceed 10 members. Several design teams may be required to complete a design in larger organisations. There are three procedural rules for the design teams (Ackoff et al. 2006):

1. Assume the unit being designed was destroyed and does not exist. Then ask the question: What system would we like to have given no system exists?

2. Equal participation. Include as many stakeholders as is feasible. Everyone is equal. There are no superior-subordinate relationships during the design process. Design facilitators will be required to ensure everyone has a say in the design and to prevent bosses from overriding their subordinates.

3. Contributions must be positive. It is important to prevent negative comments being made about proposals. If a team member does not like a proposal then he or she must propose a counter-proposal.

Ackoff et al. (2006) explain the idealised design process in detail and give examples of successful designs.

The design is compared to the mess formulation to determine the gaps that need to be filled. The gaps are used to define the ends. The key feature of this phase is *choice*: the design should increase the range of choices of all stakeholders.

Means Planning

The comparison between the mess formulation and the idealised design identifies the gaps that need to be filled by planning. Means planning determines how the gaps will be filled, e.g., by adding or eliminating new units to the SMMC. The means plan is an integrated combination of policies, practices, procedures, processes, projects and programs. It must include target dates for implementation.

The key feature of this phase is *cooperation*. Where possible, all affected stakeholders should be involved in developing the means plan, particularly those who will implement it. Greater commitment to implementing the plan is achieved through participation, and there will be less likelihood that stakeholders will oppose the plan when it is implemented if they have participated in creating it.

Resource Planning

The aim of resource planning is to determine the resources needed to implement the means plan. First, the total resources required for the means plan are calculated. Next, the resources that are available are computed. The difference between the two identifies the resource gap, the amount of resources that need to be acquired. Ways to acquire resources are devised and aggregated into a resource plan that is scheduled against the means plan. The major classes of resource that need to be considered are time, money, people, materials and components, facilities and equipment, and information, knowledge, understanding and wisdom.

The key feature of this phase is *commonweal*. This is a term not in common use today. It is intended to convey the idea that all resources are communal even though a SMMC may have privileged and legal access to them. SMMCs have temporary stewardship of resources: resource use should be considered from the viewpoint of contemporary society and future generations. Ideally resources should at least be conserved. This can be achieved by reducing the amount of energy consumed, using renewable energy, reducing waste, and recycling or remanufacturing. One company that is showing the way on resource usage is InterfaceFLOR, an environmental leader in the carpet industry. It manufactures carpet squares. Carpets are synthetic and are made from oil. Yet the environmental targets for the company are: "Zero emissions. Zero waste. Zero oil." As of January 2012 the company was halfway to achieving its targets.

Implementation and Control of the Means Plan

Executing the means plan changes the company from its current state toward the idealised design state. The key feature of this phase is *competence*. The aim is to increase stakeholder competence through learning and adaptation. A plan for controlling the implementation is developed and the implementation is monitored to ensure the changes are achieved on time and meet the targets.

Modern Socio-Technical Work Group Design

Chapter 14 presented the principles of socio-technical work groups. Social scientists have shown that it is possible to have different social systems consisting of autonomous work groups for a given technological system. The social scientists took the technological system as fixed. Working independently, production engineers developed a method of production called group technology (Chapter 4). The engineers also found that the most effective systems were people working in groups. They formed work groups by redesigning the technological system. Chapter 12 shows how to design products to facilitate group production. The product design, production engineering and socio-technical methods are complementary. They can be used to jointly re-design the technological and social systems at the same time.

Implementation follows the same steps as described in Chapter 14, but there is one key difference. The technological system is analysed and redesigned using the methods outlined in Chapters 4 and 12, before socio-technical analysis begins. Initial socio-technical analysis aims at confirming the primary work groups and their boundaries. This is followed by detailed cell design. The process is iterative with changes being made to the product, technological system and social system until a satisfactory design is achieved.

The new designs can be verified using simulation. In some cases, full-scale prototype cells may have to be created to verify designs.

Organisational Design (Structure)

Socio-technical work groups are the foundation of a SMMC. As they are largely self-managed, there is no need to have a layer of supervisors to monitor their work and the number of levels of management is reduced. The first-level managers will manage several work groups. The number of levels of management depends on the average span of control of the managers, which typically ranges from 5 to 10. Classical management theorists argue that the maximum number of subordinates a manager can manage effectively (span of control) is 8 to 10. This theoretical limit is based on two assumptions: (1) a manager directly supervises his or her subordinates, and (2) the work roles of the subordinates have a fairly high degree of interdependence. If these two assumptions are not met, then the span of control can be much larger. Table 15.2 shows the number of employees in a company versus the levels of management for average spans of control of 5 and 10.

TABLE 15.2

Number of Employees versus Levels of Management

Number of Levels	5 Subordinates per Manager	10 Subordinates per Manager
1	26	101
2	131	1011
3	656	10,111
4	3281	101,111
5	16,406	
6	82,013	

With a span of control of 10, a first-level supervisor could manage 10 work groups of 10 people, giving a total of 101 employees. A level 2 manager can manage 10 supervisors, so the total number of employees is now 1011. SMMCs may range from less than 10 employees up to several thousand. Therefore the number of levels of management could range from 1 up to a maximum of 4.

Jaques (1976, 2006) argues that there should be no more than five levels in the management hierarchy for most companies. Only very large multinational corporations have more levels. The names of the operating units at the five levels are:

I. An individual employee

II. Output team

III. Mutual recognition unit

IV. A division

V. Business unit

An output team is an autonomous work group or a staff/service team. A mutual recognition unit is the largest group of employees where they can work in close contact by mutual recognition. It contains several output teams and corresponds to a small factory or a department in a larger factory. A division is composed of several mutual recognition units. A business unit is composed of several divisions.

The aggregation of work groups to form larger operating units should be based on the work flow. If there are parallel flows, then groups performing similar work at the same stage of flow can be grouped together. For example, parallel casting groups could be grouped together in a foundry. Some groups may have to be separated because of special environmental requirements, e.g., clean room conditions, or for testing (test cell). The result is an output-based organisational structure. The operational units at each level should operate as profit centres.

Other staff will be employed who do not belong to the primary groups, but provide other essential services to the company or support to the primary groups. They include design staff, human resource staff, finance staff, technical specialists, etc. They can be divided into two structures: staff units and service units. The service units are sustaining units. They perform non-essential services to support the ongoing operations. The staff units are concerned with standardisation of the work and other expert staff roles. The output, staff and service units can be combined to form an operations/support organisation (Jaques 1976, 2006). Despite its name, it is a one-dimensional structure because it is arranged according to the outputs.

The most important aspect of Jaques' model is that operational managers at each level have the authority to set up their own supporting units. Naturally the costs of the staff and service units affect the operating units' profitability, and hence the performance of the operational managers. This provides an incentive for the managers to keep these units lean. The staff of these units report directly to the operational managers. Without this constraint staff and service units can grow uncontrollably. Figure 15.2 depicts the structure. Jaques uses standard terminology for the roles at each level.

Global markets can be reached by setting up manufacturing and technical support facilities overseas. They operate as profit centres and can be organised as described in this chapter.

The operations/support structure will suit many SMMCs as they focus on niche markets. It may create problems for those companies operating in several different markets. In this situation, it may be desirable to have a more flexible organisational structure so the company can quickly change in response to market changes. The problem is resolved by the multi-dimensional organisation (Ackoff 1981, 1999, Halal 1986). The company is divided into autonomous units along three dimensions: inputs, outputs and market. Units in the output dimension are arranged according to the outputs, the types of products. Input units are arranged according to function: they are expert staff and service units. Market units are arranged according to the markets, geography or customer type. All units should operate as profit centres.

The units interact with each other through an internal market economy under the control of an executive office. Unit accountability is defined by unit charters. The units must abide by the policies set by the executive office, but are recompensed if the policies adversely affect their profits. Units must obtain all externally acquired investment capital solely from the executive office.

The executive office is accountable for managing the determination and transmission of company values, negotiating unit charters, setting policy for all units and directing and controlling the market network. Network direction and control is achieved via the acquisition and distribution of financial capital. The executive office provides investment capital and subsidies to the units. It obtains income by charging interest on loans to units, taxing those that earn income and selling management services to the units. It should operate as a profit centre.

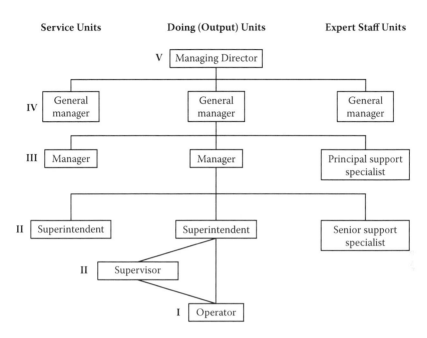

FIGURE 15.2
Operations/support organisational structure [Note: The supervisor role is not required if operators are member of socio–technical work groups.]

This type of company can change very quickly without reorganising because it is always structured along three dimensions. The units are designed so that they can be easily added or deleted. Organisational change is achieved by changing the relative proportion of resources allocated to the three dimensions.

Organisational Design (Vertical Teams)

A wide variety of teams are used in industry for development, the better-known ones being value-added management, quality circles, lean manufacturing teams, and Six Sigma teams. The names of the teams generally reflect the focus of their activities. One may wonder whether additional teams are really necessary if a company is organised into socio-technical work groups, as the groups are responsible for improvement and development. There is no need to have separate quality circles or Six Sigma teams because the autonomous work groups can undertake the tasks performed by these teams. The teams and work groups are both based on work flow. On the other hand, value-added management and lean manufacturing often require cross-functional (cross-work flow) teams to be formed.

There are two kinds of cross-functional teams: vertical and horizontal. Vertical teams facilitate democratic decision making in management. Horizontal teams facilitate multi-group and multi-functional coordination.

This section discusses how to design vertical teams. The next section discusses how to design horizontal teams.

Autonomous work groups develop team member skills in planning, decision making and problem solving. Having experienced decision-making freedom within their group, team members will expect to be involved in decision making at higher levels of management. This can be accomplished through a circular (recursive) organisation (Kobayashi 1971, Ackoff 1981, 1999). The purpose of the circular organisation is to facilitate participative planning and decision making. It is a set of boards connected to each other through the organisational structure. Each manager has a board. The members of the board are the manager, his or her subordinates and his or her immediate superior.

Consider an operational structure for a factory as depicted in Figure 15.3. The naming convention is based on Jaques' model. There are 15 work groups supervised by three superintendents. Each work group has five members. There are no supervisors as the work groups manage themselves. The superintendents report to a factory manager.

The subordinates for the superintendents' boards at level II are representatives from the primary work groups, not the whole work group. The work groups determine who the representatives are. So the board of superintendent A would have the following members:

- Superintendent A
- Factory manager
- Five representatives from the work groups reporting to superintendent A

The boards of the other superintendents are similar to that of superintendent A.

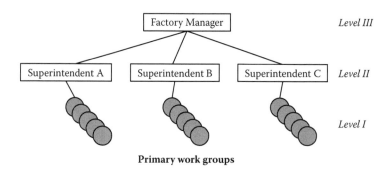

Primary work groups

FIGURE 15.3
Example organisational structure for a factory.

The board of the factory manager would have the following members:

- Factory manager
- Factory manager's superior (the CEO)
- Superintendents A, B and C

The boards for superintendent A and the factory manager are depicted in Figure 15.4.

The top-level board, normally the board of directors, should consist of the directors, the top-level manager (CEO) and his or her subordinates, external stakeholders and representatives from lower levels in the company. Staff boards can be established in a similar manner for service and expert staff.

In the example, members of the work groups interact directly with two levels of management: their superintendents and the factory manager. The superintendents interact directly with four levels: their subordinates, their own level (their peers), the level of the factory manager and the level above this. In a larger company, managers in the middle of the structure will interact directly with five levels of management. It can be seen that this is a very effective means not only for participation, but also for communication up and down the organisational structure and transmission of the company's values.

The boards can be planning boards or full decision-making boards. Planning boards plan, but the final decision for implementation lies with the manager whose board it is. The functions performed by the decision-making boards can include planning, coordination, integration, policy making,

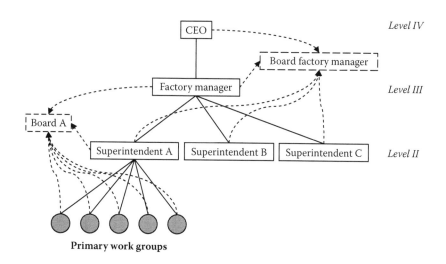

FIGURE 15.4
Boards for Supervisor A and the factory manager.

improving the quality of work life of the subordinates on the board, and evaluating the performance of the manager whose board it is. Boards should operate by consensus.

The role of the manager changes in the circular organisation. The manager is a support person for his or her subordinates. Some companies emphasise this by inverting the organisation chart. The manager's role is:

- To ensure subordinates know what their goals and tasks are, have the resources to effectively perform the tasks in their roles, and are competent to carry them out.
- Staff development.
- Promulgation of the company's values (leading).
- Decision making (budgeting, implementation of policies, etc.). Even with full decision-making boards, managers still make decisions. The boards are not an alternative to managers. They complement them.

Organisation Design (Horizontal Teams)

The autonomous work groups and circular organisation deal with day-to-day operations, and improvement and development along the work flow. Though necessary, this is often insufficient to deal with all situations. For example, different primary work groups may contain the same kind of specialist. The professional development of these specialists might be best achieved by forming them into a discipline-specific career development team. Or a problem may occur in the same type of equipment used by different work groups in different work flows. The problem may best be resolved by including people from the different groups in a problem-solving team. Or, large-scale implementation of a lean manufacturing program may require teams cutting across the work flows. We refer to teams that cut across work groups and functions as horizontal teams (Figure 15.5).

The functions of horizontal teams can include specialist development, problem solving and waste reduction. The team members will be workers from different work groups and functions. The teams should meet at least once per month. A manager is assigned to monitor team performance and to champion their proposals.

Dealing with Product Demand Variability

Product demand for batch manufacturing can vary considerably from week to week and month to month. The labour workload demand varies

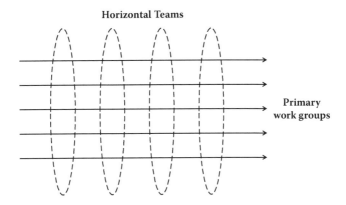

FIGURE 15.5
Horizontal teams.

accordingly. SMMCs cannot simply hire and fire full-time employees on a weekly or monthly basis to match the workload variability. Nor can they employ full-time staff based on peak workload; the labour costs will be too high. There are two very effective practices for dealing with labour workload demand variability: sub-contracting and time banks. In sub-contracting, a proportion of the peak workload is allocated to sub-contractors, with the rest performed by full-time employees. The amount of workload allocated depends on the peak-to-average ratio of the demand curve. The sub-contractors lose work first when the workload drops from the peak value. Sub-contracting reduces the workload variation that full-time employees are exposed to.

Time banks accrue overtime and undertime of full-time employees. The number of employees is set by the average workload. When workload increases above average the employees work overtime and hours are added to their time bank. When workload is below average, employees work undertime and the hours are deducted from their time bank. Employees can use the hours they have in the time bank to take leave during periods of low workload. There is a reconciliation at the end of each year. Employees with positive hours are paid the equivalent in salary and the time bank is reset to zero. If employees have negative hours—they owe hours—these are carried over to the following year.

Exercises and Problems

1. Define stakeholder.
2. List the stakeholders of a SMMC by exchange relationship. List possible stakeholders by stance.
3. List and briefly describe the phases of interactive planning.
4. A SMMC has 300 employees and 5 levels of management. Could the number of levels be reduced? Explain your answer.
5. Briefly describe the operations/support organisational structure. Discuss its advantages and disadvantages.
6. Discuss the differences between vertical and horizontal teams.

References

Ackoff, R.L. 1981. *Creating the corporate future*. New York: John Wiley.

Ackoff, R.L. 1999. *Re-creating the corporation*. New York: Oxford University Press.

Ackoff, R.L., Magidson, J., and Addison, H.J. 2006. *Idealized design*. Englewood Cliffs, NJ: Prentice Hall.

Britton, G.A. 1989. A methodology for using Beer's viable system model. *Cybernetics and Systems*, 20, 249–264.

Emery, M., and Devane, T. 1999. *Search conference*. San Francisco: Berrett-Koehler Communications.

Halal, W.E. 1986. *The new capitalism*. New York: Wiley.

Jaques, E.L. 1976. *A general theory of bureaucracy*. London: Heinemann.

Jaques, E.L. 2006. *Requisite organization*. Arlington, VA: Cason Hall and Co.

Kobayashi, S. 1971. *Creative management*. New York: American Management Association.

Mitroff, I. 1998. *Smart thinking for crazy times: The art of solving the right problems*. San Francisco: Berrett-Koehler Publishers.

Mumford, L. 1971. *The pentagon of power; the myth of the machine*. London: Secker and Warburg.

16

Organising People for Design

Introduction

The design team organisation presented in this chapter is an updated version of a model published by McCallion and Britton (1991). It is applicable to all design teams. The original study distilled the essential features for successful team performance by studying different kinds of teams in industry: quality circles, chief programmer teams, egoless programming teams, design teams, value-added management teams and planning teams. The updated version takes into account Aspelund's (2006) sub-activities of design (Chapter 3), by considering how they should be delegated to different roles.

The concept map is shown in Figure 16.1. The chapter describes the different roles in effective design teams, how design teams are linked vertically and horizontally, how they are linked into the organisational structure of a manufacturing company, and how they are managed.

The Design Team

A design team is defined by eight roles: chief designer, chief co-designer, designer, team leader, client, inventor, critic and librarian.

Chief Designer

The chief designer is responsible for client communication and management, concept development and integrity, configuration management, the quality of the design and design documentation, ensuring the design rationale and process are properly documented and technical risk management. Concept integrity is most effectively assured by making one or at most two chief designers responsible for the product architecture (Brooks 2010). Application of this rule depends on the number of levels of architecture for the product. One chief designer can design the architecture if there are a few levels.

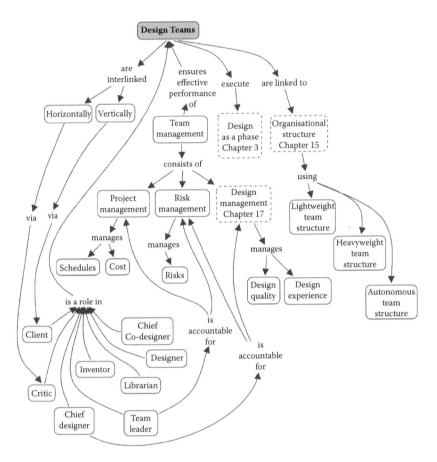

FIGURE 16.1
Concept map for Chapter 16.

The other designers support the chief designer by designing parts of the product within the architectural constraints created by the chief designer.

Several chief designers will be required if there are many levels of architecture. Each chief designer will be responsible for a few levels and is accountable to the next higher-level chief designer. Each chief designer will have his or her own design team.

The chief designer should be the most experienced designer in the team and have good interpersonal communication skills.

Chief Co-Designer

The chief co-designer is an assistant to the chief designer. The role reduces the technical risk should the chief designer leave or become incapacitated. The co-designer must work closely with the chief designer and understand the design concept and rationale underlying it. The co-designer will carry out studies for the chief designer to explore the design space and develop alternative design solutions. One person will be the chief co-designer. The chief co-designer should be an experienced designer and have good inter-personal communication skills.

Designer

The chief designer designs the concept and embodiment architectures and will also carry out detailed design if the project is small. On larger projects, the chief designer will focus solely on architectural design and the detail design will be performed by a team of designers. The designers design the components and sub-systems within the architectural constraints. The designers should be experienced in component and sub-system design.

Team Leader

The team leader is responsible for ensuring the design team finishes the design on time and within budget, providing the resources needed to com-plete the design phase activities, and freezing the design concept at various stages of the project.

The team leader role is necessary because usually designers do not have the discipline to manage schedules and budgets properly. The role allows the chief designer to focus on architectural development by relieving him or her of the project administrative and management tasks. The team leader must be an effective project manager. The team leader will be in charge of the design project, not the chief designer. In general, a person other than the chief designer should fill this role. The chief designer can be the team leader if he or she has good project management skills.

Client

The client role is a general term for all the stakeholders affected by the design, including, but not limited to, the following:

- Customer (person who pays for and uses the design)
- Secondary users (people who are affected by customers' use of the product)
- Manufacturer(s)
- Sellers/distributors
- Maintainers (people who will maintain the product)
- Disposers (people responsible for disposing of the product when it is no longer required by the user)

One of the authors *designs* his own shirts, but he is not a fashion designer. His shirts are tailor-made by a company called CYC The Custom Shop.

How can he be a designer? Isn't he the client? Well, he selects the material for the shirt, the type and size of collar, the type and size of cuffs, the style of the shirt front, the style of the shirt back, the pocket, the thread colours, the colour and size of the buttons, and additional features such as the trim on the cuffs and collar. Isn't this designing? The company website even has a feature called "Create your shirt". Note the word *create*.

Identification is one of the sub-activities of design as a personal activity (Chapter 3). Identification results in a design brief. Design briefs are generated through synthesis. They are also part of the design, because the requirements in the brief include the general functional properties of the system to be designed and may include physical features as well. In this sense, the client designs the system.

Some writers, notably Churchman (1971) and Ackoff (1976), go further and argue that the client should be the designer. The critical part of the argument relates to people's values. For a designer to design a system for a client he or she must know what the client wants, or in other words, what the client values. Unfortunately, it is extremely difficult to measure people's values. If the client's values are not properly measured, then there is no way the designer can be sure he or she is designing the right system. The problem of value measurement becomes insurmountable when designing systems with many stakeholders. One way to overcome this problem is to involve the stakeholders in the design process as designers. When this is done, there is no need to measure people's values; they will input their values as they make design judgements. The client can and should be a designer. The client should be involved in design as far as is practicable. In practice, the actual number of stakeholders could be very large. If this occurs, then representatives should be used to keep the size of the design teams manageable.

In summary, the role of the client is to create the design brief, approve the brief, design the product and approve the product design.

Inventor

The inventor is an expert designer of a sub-system or specific type of component. The inventor helps the chief designer to develop novel design solutions and resolve difficult design problems when they occur. The inventors are not full-time team members. They join the team only when required. They should be identified at the beginning of the project so they can be called in at short notice to assist.

Critic

The critic is the chief designer's enemy. The critic is responsible for evaluating the concept and whether it meets the client's needs. The interaction between the critic critiquing and the chief designer defending proves the concept. Critiquing includes system's engineering validation and verification techniques and non-quantifiable, intangible judgements, for example, judgements about political, moral, religious and aesthetic aspects of the design.

The critic must be someone other than the chief designer because of known difficulty in criticising one's self. The client may partially fill the role, but is unlikely to have the specialist knowledge to critique all aspects of the concept; therefore specialist critics are employed. There are six major methods for critiquing designs: similarity, analysis, demonstration, simulation, examination and test (see Chapter 2). The specialist critics are experts in one or more of these methods.

Librarian

The librarian is responsible for the recording, storage, retrieval and transmission of all forms of communication, including classification and coding systems. The librarian can assist the chief designer in preparing presentations for client and management communications. The librarian coordinates the communication processes between the team roles and between the team and senior management. The librarian may be one or more people and should be a specialist in communications and multimedia.

Design Team Coordination and Integration

There can be several or many design teams working on one project. The teams will be created using the role definitions above. They can be vertically integrated using a recursive organisation structure (Ackoff 1981, Beer 1979). The structure is illustrated in Figure 16.2 for a pair of teams at two levels of recursion (levels 0 and 1).

The chief designer and team leader at a higher level set the architectural and management constraints for the next lower-level team. They are clients of the lower-level team. The chief co-designer can be delegated this role if the chief designer is not available. Figure 16.2 shows the top-level team at recursion level 0 (L0) and one team at the next lower-level of recursion (level 1). Chief designer L0 and team leader L0 are clients of team 1 at L1 (team L1T1). Similarly, the lower-level chief designer and team leader are clients of the higher level: chief designer L1T1 and team leader L1T1 are clients of team L0.

Although each team is a client of the other team, there is a power asymmetry here. The lower level is subordinate to the higher one. The figure shows one pair of teams. Obviously there will be several other teams at level 1. The same pattern is repeated for these other teams.

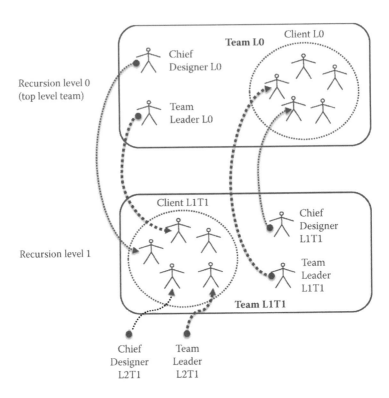

FIGURE 16.2
Vertical integration of design teams.

The figure also shows the chief designer and team leader from team 1 at the next lower level of recursion (level 2), who are clients of team L1T1 at level 1. The recursive linking allows a team in the middle to be in contact with a team one level above and one level below its level of recursion.

Horizontal coordination between teams at the same level of recursion is achieved by allowing the chief designer of one team to be a critic of the other team and vice versa, as illustrated in Figure 16.3. The chief co-designer can be delegated this role if the chief designer is not available. The pattern is repeated if there are more than two teams at the same level. If there are, say, five teams, then each team will have four chief designer critics.

Product and Manufacturing Design Team Cross-Linking

Manufacturing systems need to be modified to produce new products. A completely new manufacturing system may have to be designed for a novel product. The two design processes, product and manufacturing

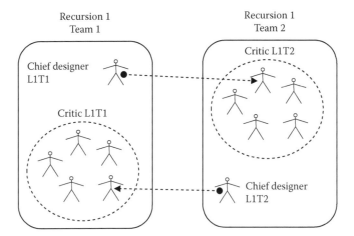

FIGURE 16.3
Horizontal coordination of design teams.

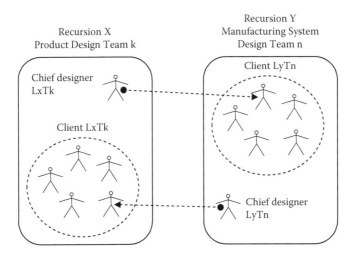

FIGURE 16.4
Product and manufacturing teams cross-linking.

system, must be harmonised and the designs integrated. Chapter 11 showed how the two life cycles can be linked together and harmonised. This section discusses how the design concepts can be integrated.

Integration is achieved through cross-linking of the design teams (Figure 16.4). The chief designer of one team becomes the client of the other team and vice versa. As clients they are involved in the design of each other's system.

The two top-level (level 0) teams will be cross-linked. A lower-level manufacturing sub-system team will be cross-linked with a product sub-system team if the manufacturing sub-system makes or assembles the product sub-system. Team linking is selective. The teams may be at different levels because the respective systems may be partitioned differently, as indicated by the level labelling in Figure 16.4.

Team Management

The major activities of team management are:

- Project management
- Risk management
- Design management

Table 16.1 shows the allocation of these activities to the team roles.

The team leader is in charge of the design project and is accountable for all project management activities, except for design management. Risk management is a joint responsibility of the team leader and the chief designer, as some risks are technical in nature and others relate to business issues. The chief designer is accountable for technical risk. The team leader is accountable for business risk. The chief designer is accountable for design management, but will delegate the administrative work to the librarian. Design management is discussed in Chapter 17.

Team management is essential to create successful designs on time and within budget. But it is not the main purpose of the design teams, which is to design a new system. There is an optimal level of management that should be imposed and an optimal level of documentation. Too little control and documentation increases the risk that something will be overlooked, resulting in poor quality or cost and schedule overrun. Too much control stifles creativity, resulting in low-quality designs. Too much documentation reduces the amount of time designers actually spend on designing and

TABLE 16.1

Allocation of Team Management Activities to Team Roles

Team Management Activity	Team Role Accountable for the Activity
Project management	Team leader
Risk management	Team leader and chief designer
Design management	Chief designer, with assistance from the librarian

increases the development costs. Whatever systems are used to manage the design projects, they should be simple, easily understood and useful.

The following sections provide a range of alternative techniques for managing teams. The art of team management is to select techniques that will be optimal in a given design situation and integrate them effectively.

Project Management

The basis of effective project management is a program that defines the following (Britton and Parker 1993):

- What is to be done (the activities to be carried out): The "what is to be done" changes during the design project as the architecture evolves. Initially it will consist of early activities to create the system architecture. Once the architecture is defined it can be partitioned into work packages for execution. The structure of the work packages is called the work breakdown structure (WBS). The WBS depends on the partitioning of the architecture and must include the additional activities to manage the design teams. The team leader and chief designer are jointly accountable for this activity.

- Who is involved (the project organisational structure): The organisational structure is defined by the roles in a design team, the horizontal and vertical interlinking between teams, and team integration with the organisational structure of the company. People are assigned to the team roles. The chief designer is accountable for recruiting designers and inventors and establishing the interlinked teams. The team leader is accountable for recruiting people to the other roles and integrating the teams with the organisational structure of the company.

- Who is to do what (the accountabilities): This is defined by a matrix of accountabilities that shows how the work packages are assigned to the teams and roles. The team leader and chief designer are jointly accountable for this activity.

- The sequencing of the activities. An activity network defines the activities that need to be carried out and their interrelationships. The team leader is accountable for this activity.

- A resourced activity network. The resourced network has resources allocated to activities and the network is verified by ensuring the resources are available when required. If not, then the network is adjusted to take into account resource constraints. For most projects the resourced activity networks can be created using critical path method (CPM) software. The resourced networks integrate costs with the schedule. The team leader is accountable for this activity.

- The rate of work (expected progress). A rate of productivity is assumed for resourced activities. It is recorded when networks are resourced. The team leader is accountable for this activity.
- The relative importance of activities. The team leader is accountable for monitoring schedules and budgets. The chief designer is accountable for monitoring the design scarce resources.
- The communication of information. The critical project information defined above must be communicated to the design teams. The team leader is accountable for communicating project management and critique information. The chief designer is accountable for communicating design information.

Project management involves planning future activities and planning and control of activities in progress. Project planning for future activities defines the design project budget, transforms the budget into resourced activity networks, determines the project organisational structure and allocates work for execution. In small projects the whole project can be planned in advance and then executed. In larger projects, project planning proceeds in stages. There is a planning horizon. The project is planned in detail up to this horizon. Activities are planned at an aggregate level beyond the horizon. As the project proceeds, the planning horizon moves forward and activities previously beyond the horizon are planned in detail. Planning proceeds like a wave moving forward.

An activity network should be established as soon as the design project commences. Long lead time items should be identified and a network created to monitor their ordering and delivery. Failure to do this will invariably result in a schedule overrun.

Planning and control of activities in progress involves the day-to-day planning of activities in progress and controls these to ensure the deliverables are achieved on time and within budget. These activities change as the project proceeds.

For more detail on how to manage projects read Britton and Parker (1993), Bent (1982), and Kerzner (2009).

Risk Management

Risk management is a systematic process for managing design project risks. It consists of four major activities: risk planning, risk identification, risk assessment, and risk mitigation and tracking (Shisko 1995, ECSS-M-00-03A

2000). *Risk planning* results in a risk plan that states the risk policy and how risk identification and characterisation, risk analysis, and risk mitigation and tracking are to be carried out. There are two basic risk policies: risk management and risk avoidance. A risk management approach accepts a certain level of risk in the project and aims to mitigate the effects. It results in an architecture with little or no redundancy. In this case, a backup redundant unit may not achieve the same level of performance as the primary unit. A risk avoidance approach, on the other hand, tries to avoid risk altogether. It leads to an architecture with 100% or more redundancy on all critical items. Backup redundant units have the same level of performance as the primary units.

Several techniques are used for *risk identification*:

- Expert interviews.
- Independent assessment.
- Learned lessons files/database: These should be kept for all projects. Good examples are NASA learned lesson files and Bechtel's project closeout reports.
- Analysis: Failure modes effects analysis, fault tree analysis, worst-case analysis and safety analysis.

Risks can be classified using EASTEEP (Section II).

Risk assessment techniques are based on probabilistic analysis. Probability models can be developed for specific decisions, assessing risk, schedules, cost and effectiveness. Using probabilities is OK as long as the probability distributions are verified by field or experimental results. If this is not done, then the distributions could be wrong and the assessment of risk will be misleading. Typical techniques used for assessment include decision analysis, probabilistic risk assessment, probabilistic network schedules (program evaluation and review technique (PERT) and graphical evaluation and review technique (GERT), and probabilistic cost and effectiveness models.

Risk mitigation and tracking techniques include the following (Shisko 1995):

- Watch list: This is a list of important risks with their early warning signs and risk mitigation strategies. The list is created and monitored. If the warning signs are detected the appropriate strategy is implemented. *Milestones* perform the same function in network schedules. The list will be updated and revised as the design progresses.
- Contingency planning: This involves foreseeing possible problems and developing solutions for them, before they actually occur. Triggering events need to be specified, which will activate the contingency plans. *De-scope* planning is a special form of contingency planning in which the solution is to reduce the scope (functionality) of the system being designed.

- Parallel development: This involves developing two competing technologies at the same time. It is used when a new technology is being developed and management is not sure that the new technology will be successful.
- Critical items list/issue log: This is a list of critical items and issues occurring during the project that could have an adverse impact on the project if they are not resolved. These are closely monitored until they are resolved.
- Integrated cost/schedule system: This is a system in which the cost systems are fully integrated with the scheduling system. Risk is reduced because the costs are always synchronised with changes in the project schedule.
- Design scarce resource and technical performance tracking: This involves monitoring the design scarce resource and key technical performance parameters against planned baselines to ensure that development is proceeding as planned. It is discussed in Chapter 17 as part of design management.
- Design strategy: Technical and business risks can be reduced by using a make-to-order or assemble-to-order strategy.

Organisational Integration

Chapter 15 described how SMMCs can be divided into operating units based on work flow. Expert and support staff are employed by the units. Usually they are divided by area of expertise or discipline. This is called a functional division of labour. The members of the design teams are drawn from the output units and the expert staff and service units. They may work full-time or part-time on the design project.

There are three ways of linking design teams with the organisation (Clark and Wheelwright 1993): lightweight, heavyweight and autonomous team structures. The authors do not agree with Clark and Wheelwright that a fourth kind of team—the functional team structure—is a viable option, because there is no chief designer to ensure concept integrity. Furthermore, the functional managers as a group cannot effectively replace the team leader role needed to ensure the project is completed on time and within budget. Watts (2012) makes a similar statement in his book when he says that committees should not lead design teams.

Lightweight Team Structure

In the lightweight team, team members remain located in their output, service or staff units under the authority of their respective managers.

Most members will work part-time on the project. A junior/middle member of management is appointed as team leader. The chief designer will usually be a middle-level designer, but not a senior designer. The team leader and chief designer are coordinators. They have no authority over the other members. Middle or senior managers directly control the team.

The lightweight team structure is suitable for small design projects using mature technology and well-established design processes. It is also suitable for training junior management and design staff to lead design projects.

Heavyweight Team Structure

In the heavyweight team a senior manager is appointed as team leader and a highly experienced designer is appointed as chief designer. Core team members work full-time on the project and are located together (co-located). Other team members will work part-time on the project as and when required. They may or may not be co-located.

The team leader has full budget accountability for the project and full authority over the team members; so does the chief designer. The functional managers are accountable for the longer-term career development of the team members because the design teams have relatively short lives.

The heavyweight team is established under a project charter specifying the purpose and objectives of the project. The charter is supplemented by a contract book that defines the organisational requirements and constraints in more detail. The teams operate under the organisation's policy and procedures.

An executive sponsor or champion is appointed to link the team to senior management. "The need is for senior management to retain the ability to guide the project and its leader, while empowering the team to lead and act. The definition of such a person is an executive sponsor. This sponsor takes on the role of coach and mentor for the heavyweight team project leader and core team, and seeks to maintain close, ongoing contact with the team's efforts" (Clark and Wheelwright 1993, p. 544).

The heavyweight team structure is suitable for larger design projects using existing technology and established design processes.

Autonomous Team Structure

In the autonomous team, team members are dedicated solely to the team and are co-located. The team leader and chief designer have total control of the team and the project. They do not have to abide by the organisation's current policy and procedures. The team is autonomous. The most well-known example is Kelly Johnson's skunk works (Rich 1989, Rich and Janos 1996).

The autonomous team is most suitable for design projects using new and untested technologies or radical design concepts.

Global Design Teams

Global design is exceptionally challenging because of differences in language, culture and design practices. Hence many companies restrict design to one location. However, there are two situations where it is advantageous to use global design teams. The first is to dramatically shorten the design lead time. A single design team might work, say, 8 hours per day. If three teams are set up in three different time zones, then it is possible to design for 24 hours per day. The design is handed from one team to the next; it is serial design. Time must be allowed each day for the handover. Typically handover takes about 30 minutes. The work should be divided so that each team can finish a complete section of design before handing on to the next team.

The second situation is to tap worldwide design expertise. Parts of the design may be contracted to specialist companies, wherever they may be located. Design is carried out concurrently.

Team management is more difficult for global design compared to local design because the design teams are not co-located. The design teams can be organised using the concepts presented earlier in the chapter. A number of conditions are essential for the teams to work effectively. A few guidelines are given below based on the authors' experience.

- Common system of units. All teams must use the same system of units.
- Common terminology. Teams must use the same technical terminology. A glossary of terms is useful for this purpose.
- Common software. All teams must use the same computer-aided design (CAD) software and the same version of the software.
- Common model. Ideally all teams should design using a common master model. Teams can access the model remotely. This saves time and money sending files from location to location and trying to keep track of the different versions. The leading CAD suppliers provide software with master model capability.
- Common design process, quality standards, and engineering standards.
- Chief designer. There must be only one chief designer in charge of all the teams. He or she has total authority for ensuring concept and architectural integrity.

- Architectural interfaces. The architectural interfaces must be clean, clearly defined and communicated to all teams.

Exercises and Problems

1. Briefly describe the eight roles of a design team.
2. Explain how design teams can be integrated.
3. Explain how design teams can be coordinated. What is the difference between coordination and integration?
4. Explain how manufacturing and product design teams can be coordinated.
5. Briefly describe the different ways design teams can be integrated with an organisational structure.
6. You are a member of a student team building a racing car to enter either Formula Student or FormulaSAE (select the competition most appropriate to your location). Specify the major sub-systems of the racing car, specify the design team roles and allocate students and sub-systems to the design team roles. You can assume all students have equivalent knowledge and experience. You must include all design team roles in your answer.

References

Ackoff, R.L. 1976. Does quality of life have to be quantified? *Operational Research Quarterly*, 27, 289–304.

Ackoff, R.L. 1981. *Creating the corporate future*. New York: John Wiley.

Aspelund, K. 2006. *The design process*. New York: Fairchild.

Beer, S. 1979. *The heart of enterprise*. Chichester, UK: John Wiley.

Bent, J.A. (1982). *Applied cost and schedule control*. New York: Marcel Dekker.

Britton, G.A., and Parker, J. 1993. An explication of the viable system model for project management. *Systems Practice*, 6(1), 21–51.

Brooks, F.P., Jr. 2010. *The design of design*. Upper Saddle River, NJ: Addison-Wesley.

Churchman, C.W. 1971. *The design of inquiring systems*. New York: Basic Books.

Clark, K.B., and Wheelright, S.C. 1993. *Managing new product and process development*. New York: The Free Press.

CYC The Custom Shop. http://www.cyccustomshop.com/cyc/Style/StyleUL.aspx (accessed April 28, 2012).

ECSS-M-00-03A. 2000. *Space project management: Risk management*. European Space Agency.

Kerzner, H. 2009. *Project management*. 10th ed. Hoboken, NJ: Wiley.

McCallion, H., and Britton, G.A. 1991. Effective management of "intellectual" teams: With specific reference to design. *Journal of Engineering Design*, 2(5), 45–53.

Rich, B.R. 1989. The skunks works' management style—It's no secret. *Aerospace*, March 8–14.

Rich, B.R., and Janos, L. 1996. *Skunkworks*. New York: Back Bay Books.

Shisko, R. 1995. *NASA systems engineering handbook*. Washington, DC: NASA.

Watts, F.B. 2012. *Engineering documentation control handbook*. 4th ed. Oxford: Elsevier.

17

Design Management

Introduction

Design is an experiential process. The stakeholders apply their expertise to the process based on past experience. During the process they gain further experience that can be used for future projects. Design management is the management of this experience: past, present and future. It includes managing client communication, the design concept, the configuration, design documentation, and other experience used and gained during the design project. This chapter discusses the different kinds of experiences that people have and how to apply and manage these during design projects. The concept map for this chapter is shown in Figure 17.1. The focus question is: What is design management? The focal concept is *design management*.

The first important classification of experience distinguishes between tacit and explicit experience (Nonaka and Takeuchi 1995). "Tacit knowledge is personal, context specific and therefore hard to formalize and communicate" (Nonaka and Takeuchi 1995, p. 59), for example, balancing while riding a bicycle. The muscular coordination required to maintain balance is internal and is extremely difficult to communicate to another person. Tacit experience is most commonly associated with physical skills, but it includes other abilities, for example, the ability to lead, be courageous, and create and manipulate mental models. Tacit experience is subjective. Explicit experience, on the other hand, can be formalised and easily transmitted to other people. Formalisation is achieved through reflection and thought. Explicit experience is objective (shared). There are four ways experience can be transmitted: tacit to tacit (socialisation), tacit to explicit (externalisation), explicit to explicit (combination) and explicit to tacit (internalisation).

Socialisation

Tacit experience is difficult to transmit using language (by definition). It is most effectively transmitted through demonstration, imitation and a structured learning environment. A structured learning environment is a sequence of learning situations of increasing difficulty. The learner is

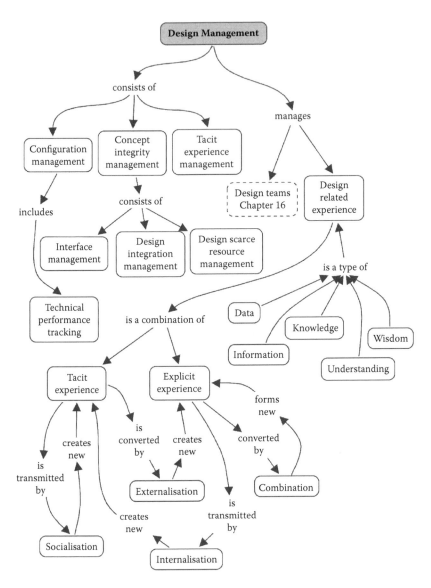

FIGURE 17.1
Concept map for Chapter 17.

guided by a skilled practitioner and is not allowed to move on to the next level of difficulty until he or she has gained mastery at the current level.

Externalisation

This is the process of articulating tacit experience into metaphors, analogies, concepts, hypotheses or models so it can be transmitted using a language.

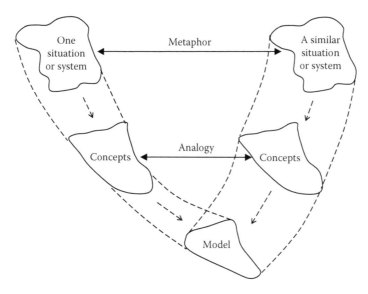

FIGURE 17.2
Externalisation through metaphor, analogy and model.

This is probably the most important form of experience creation because it converts personal, private experience into public experience that can be used by all.

Explicit experience can be efficiently created by sequential use of metaphor, analogy and model (Figure 17.2). A metaphor is an association of two different situations or systems based on one or a few similar properties. Analogy shows that two different situations or systems are similar by direct correspondence of their key features and concepts. In engineering, a formal method called the method of analogy is used to show the similarity between two different engineering models. A model is a formal generic representation of a system or situation being described or explained.

Combination

Combination is a process of organising existing ideas and concepts into a formal system. Existing explicit experience can be reconfigured through sorting (filtering), adding, combining or categorising.

Internalisation

This is a process by which a learner learns how to use and apply (internalise) explicit experience. To assist this conversion it is useful to have explicit experience "verbalized or documented in diagrams, manuals or oral stories" (Nonaka and Takeuchi 1995, p. 69). Rapid prototyping and expanding

TABLE 17.1

Different Types of Knowing

Type of Knowing	Description	Explicit Forms of Transmission	Use in Decision Making
Data	Raw observations	Measurements, test results	—
Information	Interpreted observations: know what, know when, know where, know who, know how many	Records, reports	The ability to select a course of action for a specified outcome
Knowledge	Know how (to do something)	Procedures, instructions, policies	The ability to perform a course of action (efficiency)
Understanding	Know why	Explanations	The ability to respond to changes that affect the efficiency of a course of action (adaptability)
Wisdom	Know long-term consequences of decisions	Strategies, judgements	The ability to make value judgements (effectiveness)

the scope of situations experienced also help internalisation by promoting conditions for learning.

The second important classification is based on how we use our experience to make and execute decisions (Ackoff and Emery 1972). There are five different types of knowing (Ackoff 1999): data, information, knowledge, understanding and wisdom (Table 17.1).

Data are observations, such as measurements and test results. Data are of no use to us unless they are interpreted. Data that are interpreted into a useful form are information. For example, a measurement of 100.00 kg for the weight of a microsatellite does not mean anything. First, the measurement must be performed according to international, national, industry or company standards. We need to know the standard that was used. Second, we need to trust the source of measurement: Is it an accredited or trusted source? Third, the measurement is a single measurement. This is not acceptable. A number of measurements must be taken that vary. For example, the value might be expressed as 100.00 ± 0.05. Providing we know the standard and can trust the source, then this means something. We can use the value in any situation that does not require accuracy greater than two decimal places. This is information.

Knowledge is knowing how to do something. It is the ability to perform a technique or method and is transmitted through policies, procedures and instructions. Understanding is the ability to know why something happens. It is transmitted through explanations. Wisdom is "the ability to perceive and evaluate the long-run consequences of behavior" (Ackoff 1999, p. 162). It is transmitted by heuristics, guidelines, strategies and judgements.

To learn is to acquire information, knowledge, understanding or wisdom (Ackoff 1999, p. 164). Learning can take place under constant conditions or under changing conditions (adaptation). Ackoff argues that knowledge is more important than information, understanding more important than knowledge, and wisdom more important than understanding.

Organisational experience consists of formally recorded (explicit) experience that is available to all. Companies normally focus on data and information rather than on the more important experiences: knowledge, understanding and wisdom. A large part of organisational experience consists of individual tacit experiences. Companies often overlook the importance of tacit experience. Teams and work groups are important social systems for transmitting and learning tacit experience. The next section explains how tacit experience can be managed.

Tacit Experience Management

Tacit experience is the most difficult experience to communicate to another person. Yet it is arguably the most important for design: creative synthesis is a tacit experience. Wisdom is the most difficult tacit experience to impart, followed by understanding, knowledge and information. Tacit experience affects the design process in two ways:

1. Experts share their experience with each other to design and critique.
2. Junior designers and critics need to learn how to design and critique.

Expert sharing of tacit experience can be effected through focus groups, idealised designing, brainstorming sessions and other forms of face-to-face meetings. The purpose of the meetings is to facilitate the sharing and development of ideas to solve design problems or to critique the design. Some meetings will be planned, e.g., focus groups and idealised designing. Other meetings can be ad hoc, e.g., structured walkthroughs. A structured walkthrough is a peer review of a design problem or a design solution.

Many companies ignore the tacit experience that older staff have and find out too late, when they retire, that their younger counterparts are unable to fulfil their roles to the same level of competence. Some Japanese tool manufacturers have staff who are 70 years or older still working because their tacit experience is considered to be irreplaceable. One engineering company in Singapore has recognised this issue and started a formal program to transfer the tacit experience of older staff to younger trainees before they retire.

Tacit experience transfer is best achieved through a master-trainee relationship. A staff development plan should be developed for each trainee.

The plan should consist largely of on-the-job training, but could include specialist off-site training programs. The development plan will provide a series of structured learning experiences of increasing difficulty to the trainee. The trainee is not allowed to move to the next level until mastery has been achieved at the existing level. The master both trains and guides the trainee. Bloom's and Dave's taxonomies (Chapman 2012) can be used to determine the level of difficulty and to develop the learning outcomes and an assessment framework.

Once trainees have mastered the basic skills they can be given small, low-risk projects to complete on their own under the guidance of the master. This allows them to build up design and critique experience but does not place the company at risk if they mismanage their assignments.

Design teams may not have experience in certain processes or techniques essential to complete a design. External mentors can be employed and assigned to the teams to transmit the required information, knowledge, understanding and wisdom.

Designers can also improve their knowledge and skills (tacit and explicit) through critically reflective practice.

Critically Reflective Practice

As noted in Chapter 3, design is best achieved through reflective practice (Schon 1983, Thompson and Thompson 2008). Critically reflective practice is a process by which a professional consciously considers his or her experiences in order to improve his or her professional effectiveness. It involves reflection, self-awareness and self-critique (Thompson and Thompson 2008):

- Professionals reflect on their activities as they perform them (reflection-in-action), and before (reflection-before-action) and after (reflection-on-action). Reflection-before-action is the conscious preparation for reflection-in-action. Some tools and techniques that are useful for reflection-before-action are being systematic, tree diagrams, concept maps, embedded whys (keep asking why), participative design and the RED approach (recognise conflict, evaluate conflict and deal with conflict).

- Reflection-in-action is consciously thinking while acting. Some tools and techniques that are useful are risk assessment models, reframing, noticing (paying attention), highlighting discrepancies and challenging assumptions and mental models.

- Reflection-on-action is thinking about what happened after it has happened. Tools and techniques that are useful are: What? So what?

Now what? (What happened? What is the significance of what happened? What will you do about it?), mind mapping and concept mapping.
- There are three dimensions to reflection: cognitive, affective (emotive aspects) and values (moral aspects).
- Reflection takes into account the broader organisational, social and political aspects affecting professional relationships.

Reflection is both personal and social (Thompson and Thompson 2008). During personal reflection professionals reflect privately to improve their reflective skills, time management skills, self-awareness, and clarity and focus. Dyadic reflection involves two people: the professional and a facilitator. The facilitator can be a manager, supervisor or mentor. The facilitator is a mentor, guide or support person for the professional. Group reflection is reflection in a group to improve group learning, group self-awareness, and group creativity and critiquing skills. This kind of reflection can and should be carried out by the design teams during designing.

To conclude, critically reflective practice is an integrated combination of socialisation, externalisation, combination and internalisation.

Managing Concept Integrity

We discussed how to integrate designers through interlinked design teams in Chapter 16. Here we discuss how design concepts can be integrated to produce an effective architecture. Concept integrity management consists of three main activities: interface management, design integration management and design scarce resource management.

Interface Management

There are three different kinds of interface: physical (form and fit of mating parts), functional (signal, data, fluid or energy flow) and environmental (connection through the environment). Space is nearly always a critical interface constraint. It is managed by allocating space for each sub-system, monitoring the space boundaries as the partial designs develop, and re-negotiating boundaries when required.

The interfaces are defined in an interface specification. Someone must be accountable for each interface. Problems can arise here. Interfaces often cross organisational boundaries and accountability for the interface can be avoided. The best way to resolve this is to establish design teams based on the system architecture rather than on the existing organisational structure

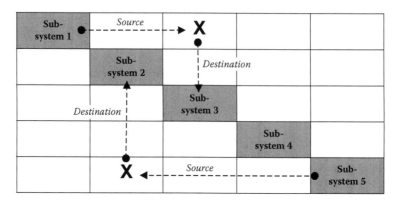

FIGURE 17.3
N² diagram format.

(Grady 1998). The heavyweight and autonomous team structures are suitable for this purpose. Interfaces can be controlled using the configuration management process described later.

Some common tools used to define interfaces are N² diagrams, schematic diagrams, computer-aided design (CAD) models or drawings, and interface dictionaries. The N² diagram (Figure 17.3) is an N × N matrix that shows the connectivity between components or sub-systems. The sub-system or component names are entered along the diagonal. The components/sub-systems can be ordered according to the architecture. A cross (X) is entered in a box that is not on the diagonal to represent an interaction (an interface). The rule for entering a cross is as follows. A cross entered above the diagonal indicates that the source of the interface is the higher sub-system (row) and the destination is the lower sub-system (column). Sub-system 1 is a source and sub-system 3 is a destination. A cross below the diagonal indicates the source is the lower sub-system (row) and the destination is the higher sub-system (column). Sub-system 5 is a source and sub-system 2 is a destination. N² diagrams can be connected together to form compound diagrams.

A schematic block diagram shows the sub-systems or components as boxes connected by lines or arrows. The connecting lines are usually labelled to uniquely identify the connection.

An interface dictionary is a table listing the interfaces, and their sources, destinations and media (Table 17.2). The dictionary is commonly used to specify data and signal connections. The number of pins in electrical connectors should be included in the dictionary under media.

The following guidelines should be used to manage interfaces (Reinertsen 1997):

- Ensure all interfaces are defined. Be careful to include backup systems and sub-systems sharing common resources.

TABLE 17.2

Interface Dictionary Format

Interface Identification		Interface Connection		
ID No.	Name	Source	Destination	Media

- Each interface should have a single point of control: one control document and one person or team responsible for the interface.
- Freeze interfaces at baselines. Have a clear procedure for including interface specifications in baseline configurations and determine when to freeze them. Freezing too early can constrain the design. Freezing too late can lead to unexpected changes in the architecture.
- Establish adequate margins at the interfaces. This constrains local changes to the local sub-systems. If no margins are allowed, then a change in one sub-system will require changes in other sub-systems.

Design Integration Management

Design is carried out in parallel by a number of different designers who produce partial designs. Ground rules need to be established before the designers start designing to ensure the partial designs are properly integrated. The key rules are:

- Common system of units. Use only one system of units.
- Common global coordinate referencing system. A global coordinate system must be specified in relation to the whole product. Designers can use local coordinate systems for their designs but must map these back to the global system.
- Common design process, quality standards and engineering standards. Common standards are desirable but not always achievable. If different standards have to be used, then a common conversion table to map the different standards to each other is essential.
- Clearly defined architectural interface. Designers are expected to work within the constraints of the interface specification.
- Common model. Ideally all teams should design using a common master model (see below).

CAD systems are used almost invariably for design. CAD software improves the productivity of individual designers and can be an effective tool for integrating partial designs. Integration is achieved using the master model concept. The partial designs and architecture are defined in

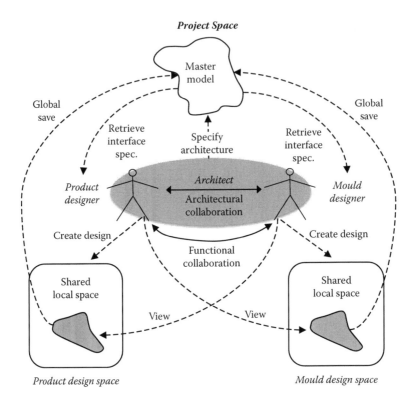

FIGURE 17.4

Collaborative design using a master model. (From Britton, G., et al., *Proceedings of the Institution of Mechanical Engineers*, 214B, 165–168, 2000.)

one and only one CAD master model. The designers have access to the one CAD model and can work on it simultaneously. Access controls prevent two designers from changing the same part at the same time. Designers can view each other's designs and thus they have an overview of how the total design is proceeding.

The master model can be used to coordinate product and manufacturing system designs. Figure 17.4 shows how plastic product design and mould design can be carried out concurrently using a master model (Britton et al. 2000). The chief mould designer is brought into the product design process early on and helps the chief product designer specify the interface between the product and the mould. This is referred to as architectural collaboration in the figure. The output of this process is the architecture that becomes part of the global, shared master model. The master model also includes other design information such as client requirements and detail designs.

Once the interface specification has been agreed upon the product and mould designers can work concurrently on their designs in their own design spaces. However, design decisions in these spaces are interlinked, and therefore the designers need to collaborate at various times during the design process (functional collaboration). The design spaces are shared in order to obtain feedback and advice on each other's design.

Concept integrity is improved if design assumptions and the rationale for design judgements (the why) are recorded with the design. First, the explanations help designers to understand and appreciate the design concept. Second, critics can check the validity of assumptions and critically evaluate judgements. Finally, at the end of the design project a review can be carried out to determine the effectiveness of the assumptions and judgements. Concept integrity can be verified using configuration matrices (Chapter 13).

Thus far we have focused on the concepts relating directly to the design. The designers also have process knowledge, that is, experience related to policies, procedures and guidelines (PPG). These should be revised and improved over time based on the project teams' experiences.

In practice this can be carried out in two main ways: during projects and between projects. The PPG can be improved during a project if it has a long duration, e.g., 1 year or more; a PPG review procedure is included in the design procedures.

PPG can be improved between projects by pooling and analysing the experiences gained from individual projects and using this as the basis for improvement. The experiences can be recorded in a project closeout report, which should be standardised so the experiences can be easily compared and analysed. The closeout report should summarise how the project was managed and identify procedural problems and successes. The following information should be recorded for the important and highlighted procedures:

- A description of how the procedure was designed and who participated in designing it
- The expected effects of the procedure
- The assumptions on which the expectations were based
- A summary of the actual information input to the procedure and the actual results
- A comparison between the actual and expected results

The report should also record all critical decisions and judgements with explanations. Lessons learned can be included in the closeout report or recorded separately.

Design Scarce Resource Management

Design scarce resource management consists of the following activities:

1. Determining the design scarce resource(s)
2. Creating a resource budget
3. Tracking resource usage against budget

4. Developing corrective action plans when the budget is exceeded

5. Managing changes to the budget

The scarce resource may be development cost (low-expense system architecture), unit cost (low-cost architecture), performance (high-performance architecture) and time to market (rapid development architecture) (Reinertsen 1997, Smith and Reinertsen 1998). The key development variables for a low-expense architecture are: maximise reuse of past designs, outsource customisation to the client (redraw the system boundary), and maximise the use of purchased parts as long as this does not compromise technological advantage. The key variables for a low-cost architecture are: reduce interface margins, design to minimise manufacturing costs, and maximise scale economies for key parts. The key variables for a high-performance architecture are: design the architecture around the critical performance-limiting subsystem or part, optimise at the system level, and reduce interface margins. The key variables for a rapid development architecture are: maximise reuse of past designs, freeze architectural interfaces early, and put design tasks with high schedule variability off the critical path.

Two points should be noted. First, reducing interface margins is equivalent to increasing the coupling between sub-systems. Second, rapid development is best achieved using mature technologies; technology development should be separated from design as discussed in Chapter 3.

Configuration Management

A product's configuration includes the design brief, design models, design critiques and other documents directly related to the design. The configuration changes in two major ways as the design evolves. First, the design brief can change (scope change). A change may be proposed by the client or by the design team. If a proposed change is significant, then a study will be conducted to determine the impact on the design, estimate the additional resources that will be required, and determine the impact on the schedule. Once the change is approved, the design budget and schedule are revised. Minor changes are usually approved without revising the budget and schedule. A scope change control process is required that identifies and records changes, ensures changes are properly approved by the client, incorporates changes in the brief, and ensures the design budget and schedule are revised where necessary.

The second kind of change is the evolution of the design and its related documentation as depicted in Figure 17.5. The process starts with an initial (first) design brief. After several iterations there is a design d and some associated documentation, e.g., stress analysis results, prototype test results, etc. There may also be several models of design d, e.g., a CAD model and physical

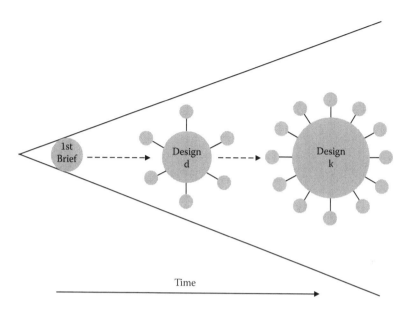

FIGURE 17.5
Evolution of design and design documentation.

prototypes. Later in the process the design is at version k and there are more documents associated with it. The different design models for each design version must be tracked and controlled. The other documents must be clearly identified with the design to which they apply. Failure to do this can have disastrous consequences. For example, assume the stress analysis for design d is erroneously believed to apply to design k because the analysis was not tracked against a particular design version. No analysis is carried out on design k, resulting in high product failure rates. A process is required to manage the design versions, to ensure all necessary documents are completed for each version, and to associate documents with the correct design version.

The changes are managed by a process called configuration management. Configuration management is a "simple, make-sense, fast, accurate, efficient, measured, and well understood process approach to planning, identifying, controlling, and tracking a product's configuration from its inception throughout its life with minimum costs" (Watts 2012). The key features of this definition are:

- Simple, makes sense: There is no point in having a complex process that most users do not fully understand or have difficulty implementing. Under these conditions they will not comply. "Makes sense" means the process must be expressed in terms that everyone can understand and be a good fit with the design and business processes. It also means there is just the right amount of control: not too much and not too little.

- Fast, accurate: The process must be quicker than the design process. A fast process ensures the configuration data and information are timely and therefore accurate.
- Well understood: The configuration management process must be well documented and the users trained to use it efficiently.
- Time frame: The process must manage the configuration until the product is retired.

Configuration management consists of two major processes: configuration specification and planning and configuration change management.

Configuration Specification and Planning

This process identifies the configuration baselines and how changes to the baselines are to be accounted for, controlled and released. The baselines are defined by the product life cycle. Usually there is one baseline for each phase and it is released after passing the phase review for the previous phase.

Each baseline should clearly define the configuration in terms of the design brief, design models (all forms of representation), the design critiques, and other documents, such as operational, maintenance and training manuals. A baseline is identified by its version number and, where appropriate, its iteration number. Each item in a configuration must be issued a number that uniquely identifies it with a baseline.

A bill of materials identifies the complete system, its sub-systems and components—parts. Parts are identified in the list by part numbers. The bill of materials is a structured list of all parts. The list is structured according to the part-parent relationships and contains other information about the parts. Bills of materials are common practice in industry because this is the data structure that enterprise resource planning (ERP) systems use. ERP systems are used by SMMCs to plan and control resources.

Many companies allow two different bills of material: one for design and the other for manufacturing. Where possible, the manufacturing system architecture should be structured to match the structure of the product architecture as far as possible (Chapter 12). Ideally, there should be one bill of material only. In general, however, the two structures may not match exactly. Configuration matrices can be used to map the product architecture to the manufacturing system architecture (Chapter 13).

Part numbering is very important. A properly planned part numbering system will be easy to understand and to update and maintain. There are three simple rules for part numbering:

1. Do not use significant digits in a part number (Watts 2012). Significant digits are codes derived from a classification system. If codes are included in the part numbers, then similar parts are found by searching the part numbers. However, if the classification system

TABLE 17.3

Illustrative Part Tabulation Table

Tab XX	Outside Diameter (mm)	Length (mm)
001	5	10
002	8	15
003	10	20

changes, then a company has to renumber all its parts! If a classification and coding system is used, and one will be if group technology is practiced as advocated in Chapter 4, then the codes can be tagged to the part numbers in a database. These codes can be searched as efficiently as searching part numbers.

2. Include a tab suffix in the part number, where appropriate (Watts 2012). A tab suffix is a number that identifies different versions of the same part. The part versions have the same geometry but vary in terms of its dimensions. The alternative dimension options are specified in the part drawing in a tabulation table and identified by tab suffixes. For example, a round pin could vary in terms of diameter and length. The tabulation table (Table 17.3) will show the preferred standard sizes of these dimensions. The suffixes might be included in the part number for the pin as follows: 3201-001, 3201-002 and 3201-003.

3. Do not renumber purchased parts.

Spares costs can be reduced by ensuring later designs are interchangeable with older designs and vice versa. An interchangeability policy and rules set the criteria to determine when parts are interchangeable. Changes can occur in form, fit and function. The best interchangeability policy is (Watts 2012, p. 74):

> Two or more items are considered interchangeable if, in all applications, they are:
>
> 1. Of acceptable form (appearance) to meet all esthetic requirements per the *product specification.*
> 2. Of a proper fit (physical) to assemble with other mating items per the *drawing dimensions and tolerances.*
> 3. Of a proper function to meet the product specification, including *performance, safety, and reliability requirements.*
> 4. These criteria must be met both ways (old design in the new and vice versa) with no special adjustments, modifications, or alterations to the items or related items.

Parts that meet some but not all of the above criteria are non-interchangeable. Watts (2012, Chapter 4) contains an extensive discussion on interchangeability illustrated with pertinent examples.

Configuration Change Management

This is a process that controls changes to the configuration baselines. It consists of the following steps:

1. Initiate a change request.
2. Classify the request and determine products and systems affected by the request (where used list). Typical change categories are: current design cannot meet requirements, cost reduction, schedule improvement, performance improvement, process optimisation and customer request (Grady 1998).
3. Evaluate the change. Conduct a study to determine the impact of the request on the design, schedule and budget.
4. Approve/reject the request.
5. Update the baseline and circulate the change and new baseline to the design teams.

Technical Performance Tracking

Technical performance tracking is essential to ensure the design will meet the performance specifications. Tracking starts as soon as the first baseline has been established. The performance measures can be generic (applicable to all sub-systems), e.g., weight or specific to certain sub-systems, e.g., optical resolution of a camera. Common generic measures include weight, volume, power, reliability, safety and maintainability. The procedure for tracking is the same as that for managing design scarce resources.

There are two methods for tracking technical performance (Shisko 1995). The first one involves preparing a planned profile for the performance measure during the design life cycle and tracking changes against this profile. For example, consider a weight constraint for a microsatellite. It is common for the weight to increase from initial estimates. Assume that the maximum weight allowed by the launch contract is 120 kg. The design weight for the first preliminary design might be set at 100 kg. Weight increases may be allowed for each of the following design reviews, with the budgeted weight at final review being 118 kg (allowing a further 2 kg margin for contingencies).

The second method involves determining the margin between the planned performance and specification and tracking changes in the margin. For the previous example, the preliminary design margin is 20 kg and the final review margin is 2 kg.

Corrective action is taken when the budgets are exceeded. In the worst case, the design may have to be de-scoped.

Exercises and Problems

1. Briefly describe the four ways design experience can be transmitted.
2. What is critically reflective practice? Does it apply to critics as well as designers? Explain your answer.
3. How is concept integrity managed?
4. What is a master model?
5. What is design scarce resource management? Why is it important?
6. How does a reduction in design margins affect the sub-system interfaces?
7. What is configuration management?
8. Briefly describe one method for tracking technical performance?

References

Ackoff, R.L. 1999. *Re-creating the corporation*. New York: Oxford University Press.

Ackoff, R.L., and Emery, F.E. 1972. *On purposeful systems*. London: Tavistock Publications.

Britton, G., Beng, T.S., and Wang, Y. 2000. Virtual concurrent product development of plastic injection moulds. *Proceedings of the Institution of Mechanical Engineers*, 214B, 165–168.

Chapman, A. Bloom's taxonomy—Learning domains. http://www.businessballs.com/bloomstaxonomyoflearningdomains.htm (accessed June 10, 2012).

Grady, J.O. 1998. *System engineering planning and enterprise identity*. Boca Raton, FL: CRC Press.

Nonaka, I., and Takeuchi, H. 1995. *The knowledge-creating company*. New York: Oxford University Press.

Reinertsen, D.G. 1997. *Managing the design factory*. New York: Free Press.

Schon, D. 1983. *The reflective practitioner*. New York: Basic Books.

Shisko, R. 1995. *NASA systems engineering handbook*. Washington, DC: NASA.

Smith, P.G., and Reinertsen, D.G. 1998. *Developing products in half the time*. New York: John Wiley.

Thompson, S., and Thompson, N. 2008. *The critically reflective practitioner*. Basingstoke, UK: Palgrave Macmillan.

Watts, F.B. 2012. *Engineering documentation control handbook*. 4th ed. Oxford: Elsevier.

Glossary

Adaptive response (to turbulence): Agree on and pursue long-term values.

Aesthetic function: The function of an aesthetic environment. Mechanical aesthetic function is perfect utility. Organic aesthetic function is perfect wholeness. Practical aesthetic function is the effect an environment has on people to produce satisfaction or dissatisfaction. Humanistic function is the ideal.

Aesthetics: A philosophy of sensuous experiences.

Analysis: A verification method that uses scientific or engineering principles to determine system performance, reliability and safety.

Animated system: A system that has purpose but its parts do not.

Assemble-to-order cycle: A cycle in which semi-finished parts are stored in inventory and then assembled when orders are received.

Autonomous team structure: A design team whose members work full-time on the design project and are co-located. The team can operate outside the company standard policies and procedures.

Autonomous work group: A work group that has significant autonomy to organise and control itself.

Bank (commercial): A financial institution that provides an efficient payment system to transfer buyers' money to sellers and collects money from depositors and packages these into loans for its borrowers.

Basic manufacturing unit (BMU): A semi-autonomous manufacturing cell. The cell consists of technological equipment and people and is self-controlled as far as possible. The purpose of a BMU is to transform materials into parts and parts into assemblies.

Batch design: Design of a product or system that will be made in small quantities for a few different customers.

Batch manufacturing: The manufacture of several different products in medium volume.

Bespoke design: Design of a product or system to meet the specific requirements of one customer.

Blank supplier unit (BSU): A basic manufacturing unit that supplies blanks of material ready for further processing.

Boundary management: Management of a socio-technical system's boundary by protecting the primary task and regulating the environmental exchanges.

Business continuity plan: A plan that identifies the anticipated risks that can temporarily stop the operations of a SMMC or disrupt its supply chain and specifies the response and response time for each threat.

Capitalism: An economic system that is based on private ownership of social and socio-technical systems to produce goods and services for profit.

Centre: An ordered volume of space that focuses our attention.

Chief co-designer: A role in a design team that is an assistant to the chief designer.

Chief designer: A role in a design team that is responsible for client communication and management, concept development and integrity, configuration management, the quality of the design and design documentation, ensuring the design rationale and process are properly documented and technical risk management.

Client: A role in a design team that refers to stakeholders who make and supply the product being designed or who are customers or users of the product being designed or other people petentially affected by the design.

Closed system: A system that has no environment.

Concept map: A visual representation of propositions. A simple proposition contains two concepts and a verb connecting them.

Configuration: A product platform and a set of variant modules to meet a particular customer order specification. It is one selection out of all possible selections.

Configuration matrix: A matrix that implements the configuration model. It is a component of the configurator.

Configuration model: The representation of product features and modules and their interrelationships that can generate valid configurations.

Configurator: A tool that supports configuration management. A configurator does not have to be based on a configuration matrix and can be manually operated or automated.

Cooperative environment: An environment containing systems that intend to help the system being designed achieve its goals and objectives.

Critic: A role in a design team that is responsible for evaluating the design concept and whether it meets the stakeholders' needs.

Defensive innovation strategy: A mixture of incremental and radical innovation. The aim is to be second to the market by developing new technologies in line with the technological leader.

Deliver-to-order cycle: A make-to-stock cycle. Finished products are made and stored in inventories ready to be sold to customers on demand.

Demonstration: A verification method that shows operational performance and compliance to requirements using the final (actual) system.

Dependent innovation strategy: Innovation from without (from external sources).

Design as a personal activity: A person designs when he or she produces a model of an object or system that does not exist, but which is capable of being brought into existence, and communicates the model to at least one other person.

Design as a phase of product development: A process that produces a model of an object or system that does not exist, but is capable of being brought into existence, and a critique of it by someone other than the designer(s).

Designer: A role in a design team that carries out detailed design.

Design management: A systematic process for managing design teams and design-related experience. It includes configuration management, concept integrity management and tacit experience management.

Design standardisation: A set of rules limiting the number of design options for the parts of a system.

Design team: A team of people, defined by roles, that creates and verifies a design.

Design-to-order cycle: The cycle is from design to deployment. Technology development is carried out separately from the order cycle.

Deterministic system: A system that has no purpose and neither does its parts.

Dissociation: An individual maladaptive response to a turbulent environment in which a person withdraws from social interactions. This response has no direct impact on turbulence, but only indirectly through superficiality and segmentation.

Disturbed reactive environment: A placid clustered environment that contains more than one system of the same kind. The interactions between the same systems create feedback loops between the parts. The background is stable.

Ecological system: A closed system that has no purpose, but has some parts that are purposeful systems.

Economic system: An abstract system that describes and explains how scarce resources are or should be allocated in a society.

Embodiment or physical architecture: An architecture that shows the major physical sub-systems and their interconnections.

Engineer-to-order cycle: A cycle in which all activities from technology development to deployment are carried out.

Environmental aesthetic: A branch of aesthetics that studies the aesthetic appreciation of all human environments—natural ones and those created by people.

Environmental morality: A morality that loves the environment as a significant other and aims to look after it.

Environment of a system: Elements and their relationships that are not part of the system or its boundary, but which affect or are affected by the system.

Evolutionary life cycle: An iterative and incremental life cycle in which all phases are repeated.

Examination: A verification method that determines the physical characteristics by visual inspection or measurement.

Final assembly unit (FAU): A basic manufacturing unit that assembles modules into final products.

Fractional reserve banking: A banking system in which each bank is allowed to keep a proportion of its deposits and lend out the rest.

Function or concept architecture: An architecture that shows the major functions or concepts and their interrelationships or interconnections.

Group technology: A methodology for designing, implementing and operating manufacturing cells based on families of parts.

Heavyweight team structure: A design team that has full-time core members who are co-located during the design project. The team operates under company standard policies and procedures.

Heuristic architectural design methodology: One that uses experience-based guidelines.

Horizontal teams: Teams in a manufacturing unit or company whose members are from different functional or operating units at the same organisational level.

Hostile environment: An environment containing systems whose intention is to prevent the system being designed from achieving its goals and objectives.

Ideal: An outcome that cannot be attained but can be approached indefinitely.

Idealised design: A design of a SMMC or social system that the stakeholders would like to have. The design is intended to help the stakeholders pursue ideals.

Ideal state of beauty: *See* ideal state of re-creation.

Ideal state of cooperation: An ideal that has the societal function to improve cooperation among people to effectively develop and utilise earth's finite resources.

Ideal state of the good: An ideal that has the societal functions to remove conflict within and between people, and for people to live in harmony with nature.

Ideal state of nurturance: An ideal that has the societal function to seek knowledge and understanding to nurture nature and humankind.

Ideal state of omnicompetence: An overriding ideal in which everyone has the unlimited ability to achieve their desires. It consists of four sub-ideals: ideal state of cooperation, ideal state of good, ideal state of re-creation and ideal state of nurturance.

Ideal state of plenty: *See* ideal state of cooperation.

Ideal state of re-creation: An ideal that has the societal functions of intensifying and purging emotions.

Ideal state of truth: *See* ideal state of nurturance.

Imitative innovation strategy: Incremental innovation that copies other companies' technologies and improves them.

Incremental life cycle: A life cycle in which the product is delivered in several versions (increments), with each version delivering more functionality than the previous one.

Induced environment: An environment that is created because a system exists.

Innovation: An invention that has been commercially exploited. It is technically and commercially feasible and socially desirable.

Interactive planning: A continuous planning process of a SMMC or social system that redesigns the SMMC or social system and implements actions to realise the design.

Invention: A new idea, concept, product or process or new modification to existing products and processes. It is something that is new and technically feasible.

Inventor: A role in a design team for someone who is an expert designer of a sub-system or a specific type of component.

Iterative life cycle: A life cycle where phases of the life cycle are repeated several times for the whole product.

Jobbing manufacturing: The manufacture of many different products in very low volumes.

L_{11}: Interdependencies between elements contained wholly within the system. These define the system's characteristic way of behaving.

L_{12}: Interdependency produced by a process that starts in the system but ends in the environment; i.e., it crosses the system boundary from the inside. It defines what the system can do to the environment. This kind of interdependency is referred to as an output in engineering.

L_{21}: Interdependency produced by a process that starts in the environment but ends in the system; i.e., it crosses the system boundary from the outside. It defines what the environment can do to the system, i.e., the system's responsiveness to changes in the environment. In engineering, this kind of interdependency is referred to as an input.

L_{22}: Elements or processes that are contained wholly within the environment. These affect the system indirectly through the L_{12} and L_{21} processes.

Librarian: A role in a design team that is responsible for the recording, storage, retrieval and transmission of all forms of communication, including classification and coding systems.

Life: Life An emotive relationship we have with the space of each object and system in our environment. Every object and system has life.

Life cycle: A model of a repeatable process, which is the complete life cycle of a system.

Lightweight team structure: A design team in which team members remain located in their output, service or staff units under the authority of their respective managers.

Make-to-order cycle: A cycle in which a standard design is created and products are then made to order.

Mass design: Design of a product or system that will be made in very large quantities for a large number of different customers.

Mass manufacturing: High-volume manufacturing of a few products.

Matrix of variances: A matrix showing the key process variances and their interrelationships.

Modularisation: The decomposition of a product into modules (building blocks) with clearly specified interfaces.

Modular product architecture: The common platform modules and their features, the variant modules and their features, and the interrelationships between these.

Modular product family: A product family in which product variants are formed by the substitution/addition of modules.

Module: A set of parts that are designed to achieve a specific function. A module can consist of one part only. It can be an assembly but need not be. Modules must have clearly defined interfaces.

Module assembly unit (MAU): A basic manufacturing unit that collects parts into sets or assembles them into modules. Modules may be assemblies but this is not essential. For example, a styling module may consist of several parts that are not directly physically connected. A MAU that "assembles" these kinds of modules is acting as a marshalling store to accumulate parts and package them into part sets (the modules) ready for assembly.

Money: A medium of exchange, a unit of account and a store of value.

Moral hazard: A situation where a person takes risks that he or she would normally avoid because there are no penalties or negative consequences for failure, but there are very high incentives if the risk works in that person's favour.

Natural environment: An environment defined by physical and biological properties.

Negative aesthetic experience: One in which a creative or heroic mood or catharsis is not achieved and the resulting experience reduces the observer's ability to create aesthetic experiences.

Neutral aesthetic experience: One in which a creative or heroic mood or catharsis is not achieved and there is no impact on the observer's ability to create aesthetic experiences.

Non-cooperative environment: An environment containing systems that are not intending to prevent the system being designed from achieving its goals and objectives, but do so unintentionally.

Normative architectural design methodology: One that codifies past experience for future use.

Offensive innovation strategy: Radical innovation that gives technical and market leadership.

Open systems planning: Planning for a socio-technical system based on an exploration of its environment.

Operations/support organisational structure: An organisation structure that is sub-divided according to operations and support structures are designed to match and align with the operational structure.

Opportunistic innovation strategy: Niche innovation leading to market leadership.

Parallel life cycle: A life cycle in which the phases are completed in sequence but are allowed to overlap.

Participative architectural design methodology: One that involves some or all of the stakeholders in the design process.

Part manufacturing unit (PMU): A basic manufacturing unit that makes parts.

Pattern architectural design methodology: One that creates a partial architectural design that can be used over and over again in different situations.

Placid clustered environment: A static environment that has serial connections between some of its parts.

Placid random environment: A simple, static environment with no connections between its parts.

Political system: An abstract system that describes and explains how a society is or should be governed.

Pollution: An act by one or more persons that adversely affects other people or the environment.

Positive aesthetic experience: One in which a creative or heroic mood or catharsis is achieved.

Product family: A set of products or product variants that share some common components or sub-systems but also contain different components and sub-systems to meet different customer requirements.

Product platform: The set of common components/sub-systems that is contained in all products in a product family.

Profession: A vocation that has the following features:
- Mastery of a body of knowledge and skills that take a long time to learn.
- The knowledge and skills provide an important service to society.
- A social contract between the profession and society.

Project management: A systematic process for managing projects.

Project manufacturing: One-off manufacture usually in a fixed position layout.

Purposeful system: An open system that can (1) produce the same function in the same environment using different courses of action and (2) produce different functions in the same and different environments.

Rational architectural design methodology: One that uses science and technology to develop rules and methods for designing architectures.

Reflective journal: A written record of your reflections as you activities. A learning reflective journal records your reflections as you learn during your study. A professional reflective journal records your reflections on your professional practice.

Regenerative development: A sustainable development approach that aims at conserving and nurturing the environment for future generations of people as well as the current generation.

Risk management: A systematic process for managing design project risks.

Role: A set of expectations about how a person is to behave in a social group or company.

Scale-based product architecture: The platform and scaling design features and their interrelationships.

Scale-based product family: A product family in which product variants are formed through different values of scaling variables.

Segmentation: A group maladaptive response to a turbulent environment in which the total social field is divided into smaller fields (segments). Each person restricts the number of people he or she will interact with.

Similarity: A verification method that shows a current design is similar to a past design that has already been qualified.

Simulation: A verification method that uses a prototype of a system to determine system performance or compliance to requirements.

Small and medium manufacturing company: A company that designs and makes its own products.

Social system: A system that has purpose and parts that also have purposes.

Stakeholder: A person or social system that is directly or indirectly affected by a SMMC.

Statement of purpose: A statement that defines the desired identity of a SMMC or social system.

Superficiality: A group maladaptive response to a turbulent environment in which the deeper roots of humanity are denied.

System: A set of elements directly or indirectly related to each other.

System architecture: A system structure designed by people.

System boundary: A volume of space surrounding an open system that separates it from its environment.

System life cycle: A model of a repeatable process that is a complete life cycle of a system.

System structure: The pattern of relationships between the elements.

Table of variance control: A table showing the key process variances and how they are controlled.

Team leader (design): A role in a design team that is responsible for ensuring the design team finishes the design on time and within budget, providing the resources needed to complete the design phase activities, and freezing the design concept at various stages of the project.

Test: A verification method whereby the performance of a system is determined under stimulated or simulated environments.

Test-to-fail: A test in which test items are tested until they fail. This type of test is used to measure reliability and durability.

Test-to-pass: A test in which a test item is expected to perform its functions under normal environmental and operating conditions.

Traditional innovation strategy: Minimum innovation.

Turbulent environment: A dynamic environment where the dynamics is created independently of systems of the same kind. This kind of environment is extremely complex and highly unpredictable.

Unfolding: An incremental, evolutionary design process to design the space structure of a system.

Vertical teams: Teams in a manufacturing unit or company whose members are from different levels in the organisational structure. The most effective design is recursive teams: the circular organisation.

Wandering: A human-powered activity performed solely for aesthetic appreciation.

Waterfall life cycle: A life cycle in which the phases are completed in sequence with no overlapping.

Wilful blindness: A situation in which a person pleads innocence on the basis that he or she did not know what was going on, but he or she could and should have known.

Work: An agreement between two or more people, in which one party agrees to perform stated tasks for the other party.

Work role: The set of behavioural expectations associated with the agreement to perform tasks.

Appendix A: Reliability and Safety Methods

Reliability and safety analyses are performed to ensure products and technical systems will function reliably and safely during their design lifetime. This appendix briefly describes five reliability and safety techniques. Readers interested in furthering their knowledge in this area should refer to Ericson (2005), which covers a very wide range of techniques.

Failure Modes, Effects and Criticality Analysis (FMECA)

The aim of FMECA is to identify the failure modes and assess their consequences. It covers all failure modes, including component failures. It does not detect design deficiencies, but does identify the consequences of deficiencies if they should occur. There are two main types of FMECA: design and process (manufacturing process). Analysis is usually performed by multi-disciplinary expert teams as the identification of modes requires expert knowledge and experience.

The basic procedure for carrying out FMECA is:

1. Define the product.
2. Prepare the functional and reliability block diagrams.
3. Identify the potential failures.
4. Evaluate each failure mode and assign a severity category.
5. Assess the probability of occurrence.
6. Identify the failure detection methods and compensating provisions.
7. For all critical items, identify corrective design or other actions to eliminate, mitigate or control risk.
8. Document the analysis.

For more details refer to Ericson (2005) and ECSS-Q-30-02A (2001).

Worst-Case Analysis (WCA)

The aim of this kind of analysis is to identify limits of performance, assess risk and improve the design. It determines whether the design will function properly and whether components will be degraded or damaged under expected worst-case conditions, i.e., when the design values reach extreme values. Figure A.1 shows the application area of WCA compared to functional performance analysis. It is assumed the system is used properly and there are no component/system defects. The justification for this kind of analysis is that although the probabilities are low for the extreme values, the consequences and cost of failure can be very high.

Three sub-analyses are performed as part of WCA: functions margins analysis, stress margins analysis, and applications analysis. *Functions margins analysis* involves the calculation of minimum or maximum values of each system function. These values are then compared with the specifications to determine whether they are within the limits. If a value falls outside the limits, then the out-of-specification value is highlighted as an alert case. *Stress margins analysis* involves the calculation of the maximum applied stress for each stress property of each component. The values are then compared with the design allowable limit and out-of-specification values are highlighted as alert cases. *Applications analysis* involves the evaluation of other application data to identify design constraints not documented in the primary specifications. Application violations are highlighted as alert cases.

The inputs to the analysis are the design specifications, the design variables (controllable variables) and the design equation, which relates the design variables to the performance output. Three performance values are normally calculated: minimum, maximum and average. The procedure can also be improved by calculating several output values over the complete range of

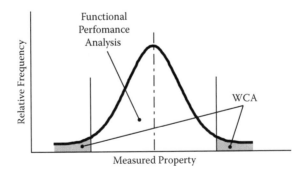

FIGURE A.1
Application of WCA.

the design variables; normally around 100 values is sufficient. This produces a histogram that can be used to determine alert probabilities and optimise the design.

For more details refer to Walker (1998).

Fault Tree Analysis (FTA)

The aim of FTA is to determine the probability of occurrence of specific critical failures. The starting point is a set of unacceptable failure modes that have already been identified. The design is then examined to determine the combination of conditions that will produce each failure mode. The combinations are represented in a tree diagram—hence the name. The tree is developed using Boolean logic. A probability of failure is associated with each branch and leaf in the tree. Therefore using the tree, it is possible to calculate the probability of occurrence of a particular failure mode.

For more details refer to Ericson (2005).

Sneak Analysis

Sneak analysis is the only analysis method that aims at finding design errors. It applies to switched circuits, that is, circuits that have alternative paths activated by switching mechanisms, for example:

- Electrical/electronic circuits
- Mechanical circuits
  - Hydraulic
  - Pneumatic
- Batch process chemical plant
- Computer software

It can also be applied to procedures and processes.

What is a sneak? A *sneak* is a latent design condition that may cause an unwanted event to occur or may prevent a desired event from occurring and is *not* caused by component failure. That is, a sneak occurs when all components are functioning normally. Sneak analysis looks at all possible combinations in a switching circuit to determine the output events and compares these to the desired design conditions. When anomalies are found the circuit is re-designed to prevent the unwanted events from occurring. The different types of sneak are:

- Sneak path
- Sneak timing
- Sneak indication
- Sneak label

A *sneak path* is an unintended flow of energy, mass or data along a circuit path where it is not supposed to be. For example, a valve that is left open, when it should be shut, may allow fluid to flow from one part of a circuit to another and cause an explosion due to mixing of different chemicals.

A *sneak timing* occurs when a flow of energy or data either activates a component or system or prevents a component or system from activating when it is not expected. An example the authors have personally experienced occurred in a lift. The sequence is as follows. Person A on the fourth floor, say, presses the lift button to go down. The lift arrives at the fourth floor and Person A steps inside the lift. Before Person A selects the floor he or she wants to go to, Person B on a higher level, say floor 10, presses the button for a lift. Person A then selects the desired floor number (first floor). The lift goes up to level 10 and not down to level 1. This is sneak timing and occurs because the lift designer has not allowed enough delay time in the circuit sequence for people to enter the lift and select the desired floor number.

A *sneak indication* occurs when instruments give a false or ambiguous reading. A sneak indication occurred during the Three Mile Island incident. The technicians were led to believe a circuit was closed when in fact it was open, due to a faulty panel light indication.

A *sneak label* is one that initiates the wrong response in a person. Figure A.2 shows a proposed design for the dials in an instrument panel. It is common practice to align the dials such that the normal positions are all oriented in the same direction. An operator does not have to read the dials, but merely look at the pattern. In the proposed design the normal position of the dials is shown by the arrows. It can be seen that the dials are not aligned in the same direction and could lead to a person acting incorrectly, by assuming something is wrong when it is not, or vice versa.

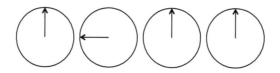

FIGURE A.2
Example of a sneak label.

Typical techniques used to conduct a sneak analysis include systematic inspection by experts, fault tree analysis, topographical analysis and Boolean algebra.

For more details refer to ECSS-Q-40-04A Parts 1 and 2 (1997a, 1997b), NASA (1999), and Ericson (2005).

Safety Analysis

The aim of a safety analysis is to ensure the safety of people operating or maintaining the system being designed. Safety analysis is a systematic procedure for determining the potential hazards that could be produced by a system and the actions that should be taken to remove or mitigate them. Analysis is usually performed by multi-disciplinary expert teams, as the identification of hazards requires expert knowledge and experience.

Each type of equipment and industry has its own safety standards; thus different procedures have been developed. The reader should refer to the relevant industry standards. One widely used method is called HAZOPS (hazard and operability study).

For more details refer to Ericson (2005) and O'Connor (2001).

References

ECSS-Q-30-02A. 2001. *Failure modes, effects and criticality analysis (FMECA)*. Noordwijk, The Netherlands: ESA Publications Division.

ECSS-Q-40-04A. 1997a. *Part 1: Sneak analysis—Part 1: Method and procedure*. Noordwijk, The Netherlands: ESA Publications Division.

ECSS-Q-40-04A. 1997b. *Part 2: Sneak analysis—Part 2: Clue list*. Noordwijk, The Netherlands: ESA Publications Division.

Ericson, C.A., II. 2005. *Hazard analysis techniques for system safety*. Hoboken, NJ: Wiley-Interscience.

NASA. 1999. *Public lessons learned entry 0756: Sneak circuit analysis for electro-mechanical systems*. http://www.nasa.gov/offices/oce/llis/0756.html (last accessed June 18, 2011).

O'Connor, P.D. 2001. *Test engineering*. Chichester: John Wiley.

Walker, N.E. 1998. *The design analysis handbook: A practical guide to design validation*. Boston: Newnes.

Appendix B: Testing

Product Testing

The objectives of product testing are to validate and verify:

- Requirements
- A design
- Manufacturing
- Installation/commissioning
- Operations
- Maintenance and support functions
- Phaseout
- Compliance with regulations or standards

There are three main types of testing:

1. Functional performance testing
2. Reliability and durability testing
3. Contractual, safety and regulatory testing

There are two main test approaches: test-to-pass and test-to-fail. *Test-to-pass* is used to show compliance with specifications. It includes functional performance testing, design verification testing, operational life testing and periodic product qualification. *Test-to-fail* is used to provide assurance of strength, reliability and durability. It includes accelerated stress testing.

Functional performance testing involves tests that simulate the normal environmental and operational conditions a product will be exposed to during its lifetime. The aim is to demonstrate that the system will function to the required performance standards. The item being tested either passes or fails the test. In general there are two types of test limits: qualification and acceptance. The acceptance test limits are based on the maximum expected operating conditions. The aim is to pass the items on test without degrading the items being tested, as they will be used by the customers. This type of testing is carried out once the design has been finalised and the products are being manufactured for use. The qualification test limits are set outside the acceptance test limits. The aim is to make sure there is sufficient safety margin in the design. Qualification testing is carried out during design and the results are used to improve the design. Design prototypes are developed specifically for these tests.

Environmental tests simulate worst-case environmental conditions a product will be exposed to during its lifetime. The environment is defined by an environmental specification. It is normal practice to use existing standards for the environmental specification and existing test procedures (to save test costs), and then tailor these to correspond to the actual conditions to which the system is expected to be exposed. The tailoring process determines both the design requirements and the test procedures for environmental testing of the actual system. The number of failures during testing is expected to be less than the specified requirements. Reliability and durability are calculated from the test results. It is assumed that reliability and durability can be measured and determined in a manner similar to that for other functional performance properties. The types of environmental tests that need to be carried out vary from industry to industry.

Accelerated stress testing (AST) is different from other types of testing. In this case, the objective is to improve design robustness and to prevent/minimise infant mortalities (early product failures in the field). The aim of testing is to stimulate latent defects so they can be detected before the product is shipped to the customer. The test conditions do not simulate expected environmental or operating conditions. They are set to stimulate failures (from latent defects); therefore the stress levels are always well outside industry standards for qualification limits. For more details on AST refer to Chan and Englert (2001), Hobbs (2000) and McClean (2000).

For a general introduction to product testing refer to O'Connor (2001).

Manufacturing Process Testing

Manufacturing processes are qualified by determining and controlling the process capability. Process capability is defined as the ability of a process to satisfy customer requirements. The most common quantitative definition of process capability is the process spread, or 6σ, when the process is statistically stable.

Control charts are used to ensure that a process is statistically stable. Two main types of control charts are used for variables measurement: X bar chart and R (or s) chart. The X bar chart controls the process mean and the R chart controls the variance. Control charts are based on sampling distributions. They plot and track the statistics of samples of parts, not individual parts. The X bar tracks the sample means and the R chart the sample ranges.

There are over 60 different measures of capability (Bothe 1997), which can be classified into short- or long-term measures and potential or performance measures. The difference between short- and long-term variation is short-term capability always has smaller variance than long-term capability, the difference being due to a shift in the process mean or an increase in variance. A process has potential capability if it can meet the minimum requirements of process capability by shifting the mean. It has performance capability if it is actually meeting the minimum requirements (that is, the mean is correctly positioned relative to the design requirement limits). In practice, companies use more than one measure in order to properly assess a process.

When the manufacturing processes are proven to be capable they can be used to manufacture products. After manufacturing, product testing is often carried out to find and remove faulty products. This kind of testing is called *screening*. It is possible to use AST principles for stress screening. *Manufacturing stress screening* is the application of stresses that will cause defective production items that pass other tests to fail on test. It identifies *weak* products so they can be removed. Normally testing is carried out on all the products. The testing is only carried out to find manufacturing defects; it is not carried out to find design defects. Note also that screening must not damage or significantly reduce the life of good products.

References

Bothe, D.R. 1997. *Measuring process capability*. New York: McGraw-Hill.

Chan, H.A., and Englert, P.J. (eds.). 2001. *Accelerated stress testing handbook*. New York: IEEE Press.

Hobbs, G.K. 2000. *Accelerated reliability engineering*. Chichester: John Wiley.

McClean, H.W. 2000. *Halt, Hass and Hasa explained: Accelerated reliability techniques*. Milwaukee, Wisconsin: American Society for Quality.

O'Connor, P.D. 2001. *Test engineering*. Chichester: John Wiley.

Appendix C: Phase and Review Definitions for Product Manufacturing System Intersecting Cycles

This appendix lists the six phase and review definitions developed for the intersecting cycles shown in Figure 11.11, which is repeated below.

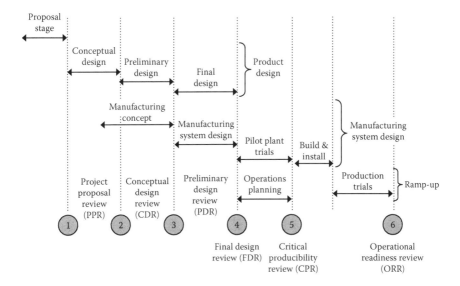

Proposal Phase

Goal: To generate and obtain concurrence on the product development project proposal.

Description: Quantification and articulation of customer needs, development of the project proposal and generation of concurrence on the proposal.

Inputs:

- Market strategy.
- Source: Marketing department.

- Technology strategy.
- Source: Chief technical officer.

Tasks and accountabilities matrix:

TABLE C.1

Tasks and Accountabilities Matrix for Proposal Phase

Task	Functions and Roles Accountable for the Task
Develop customer requirements, product specification, product concept	All major functions covering the life cycle; one person must be appointed as the project manager
Competitor analysis	Marketing
Project portfolio analysis	Marketing, chief technical officer, project manager
Develop preliminary project plan and costing	Project manager

Useful tools:

- Brainstorming
- Delphi technique
- Consumer idealised design
- Mind mapping
- Market segmentation analysis
- Quality function deployment (QFD)
- Standard proposal template
- Standard project template

Deliverables: A project proposal containing the following:

- Executive summary
- Customer requirements
- Product specification
- Product concept
- Project costing (for the next phase)
- Competitor analysis
- Project portfolio analysis (fit with the market and technology strategies and current project portfolio)
- Configuration management plan
- Preliminary project plan
  - Timeline
  - Milestones for the next phase
- Signatories

Project Proposal Review (PPR)

Description: A review to permit further investigation into the proposal by allowing the proposal to proceed to the conceptual design phase. The review assesses the proposal in terms of its fit with the market and technology strategies and the current project portfolio. If the proposal is accepted, the customer requirements and product specification are frozen.

Membership:

- Senior management
- Executive champion (if appropriate)
- Chief technical officer
- Project manager
- Members of the team who prepared the proposal
- Someone to record the meeting

Success criteria: Success criteria include affirmative answers to the following exit questions:

- Is the proposal sufficiently detailed and understood to allow it to proceed to the conceptual design phase?
- Is the proposal compatible with the market strategy?
- Is the proposal compatible with the technology strategy?
- Does the proposal fit with the current project portfolio?
- Is funding available for the next phase?
- Are resources available to complete the next phase within the proposed schedule?
- Is there concurrence on the proposal?

Documentation: Record the review panel decision and get all members of the panel to sign off.

Conceptual Design Phase

Goal: To generate a functional architecture of the product.

Description: To finalise requirements, develop the functional architecture, demonstrate technical and manufacturing feasibility and prepare preliminary risk, quality and project plans.

Inputs:
- Project proposal.
- Source: Proposal review panel.
- External and internal standards.
- Source: Chief technical officer.

Tasks and accountabilities matrix:

TABLE C.2

Tasks and Accountabilities Matrix for Conceptual Design Phase

Task	Team Role Accountable for the Task
Develop project plan and costing	Project manager
Prepare product costing	Chief technical officer
Develop quality plan	Chief designers (product and manufacturing system)
Develop risk plan	Project manager and the chief designers
Develop design concept	Chief designer (product)
Prepare preliminary manufacturing plan	Chief designer (manufacturing system)
Prepare critique plan	Chief critic

Useful tools:
- Functional analysis, functional flowchart
- Modular function deployment (MFD)
- Modelling tools for industrial design
- Timescaled Gantt chart
- Checklists and templates

Deliverables:
- Design concept
  - Functional architecture (product functional configuration baseline)
  - Product positioning (A, B or C product)
  - Material selection
  - Industrial design
  - Proof of design concept feasibility, including engineering models
- Preliminary manufacturing plan
  - Rough-level make-or-buy analysis
  - Rough-level work flow design
- Product and project costing
- Risk plan

- Quality plan
- Critique plan
  - Model/prototyping plan
  - Test plan
  - Critique (verification) matrix
- Project plan
  - Version plan
  - Timeline
  - Rough resource allocation
  - Milestones for the next phase

Conceptual Design Review (CDR)

Description: A review to accept or reject the project proposal. At this stage, there should be sufficient technical information available to assess the company's capability to execute the project, the risk associated with the project and the financial and market consequences of proceeding with the project.

Membership:

- Senior management.
- Executive champion (if appropriate).
- Chief technical officer.
- Project manager.
- Chief designers.
- Chief critic.
- External expert critic: It is useful to seek external independent critique of the concept.
- Someone to record the action items.

Success criteria: Success criteria include affirmative answers to the following exit questions:

- Is the proposal sufficiently detailed and understood so that a decision to accept or reject can be made?
- Does the company have the technical capability to complete the project successfully?
- Are the risks known and manageable?
- Can the project be completed within the proposed timeline?
- Are funds available?

- Is the product likely to be a market success?
- Are the quality and critiquing plans adequate for the project to succeed?

Documentation: Record the review panel decision and get all members of the panel to sign off.

Preliminary Design and Manufacturing Concept Phase

Goal: To generate the product preliminary design integrated with the manufacturing concept.

Description: Design the physical architecture, sub-unit design, design verification plan and the integration and development of the manufacturing concept.

Inputs:

- Conceptual design phase deliverables.
- Source: Conceptual design review team.

Tasks and accountabilities matrix:

TABLE C.3

Tasks and Accountabilities Matrix for Preliminary Design and Manufacturing Concept Phase

Task	Team Role Accountable for the Task
Project management	Project manager, team leaders for the product and manufacturing system design teams
Quality management	Chief designers for the product and manufacturing system design teams
Risk management	Team leader and the chief designers
Design management	Chief designers and librarians
Develop marketing plan and sales configuration matrix	Marketing
Design physical product architecture, construct qualification models and prototypes	Chief designer (product), client
Sub-unit design	Designers (product), client
Develop manufacturing concept	Chief designer (manufacturing system), client
Critique	Critics

Useful tools:

- Modelling tools for design (e.g., computer-aided design (CAD), physical mock-ups, etc.)
- Critiquing tools

- Finite element analysis (FEA)
- Failure mode effects analysis (FMEA)
- Worst-case analysis
- Fault tree analysis
- Sneak analysis
- Safety analysis
- Risk management tools
  - Contingency planning
  - De-scope planning
  - Watch lists
  - Critical items and issue list
- Modelling tools for industrial design
- Timescaled Gantt chart
- Checklists and templates
- Flow process charts

Deliverables:

- Detailed industrial design
- Qualification models
- Assembly model or drawings (preliminary design configuration baseline, i.e., product structure)
- Sales configuration matrix
- Preliminary marketing plan
- Sub-unit models or drawings
- Design critique documentation
- Product integration plan
- Product service concept
  - Maintainability
  - Upgradability
- Updated risk plan
- Updated quality plan
- Updated critique plan
- Manufacturing concept
  - Manufacturing functional configuration baseline (i.e., process flow description)
  - Preliminary methods selection
  - Throughput time analysis
- Time analysis

- Throughput analysis (production rate)
  – Manufacturing target costing
  – Definition and selection of critical manufacturing equipment
  – Preliminary maintenance concept
  – Preliminary manufacturing quality plan
  – Preliminary manufacturing capital estimate

Preliminary Design Review (PDR)

Description: A review to ensure product design can proceed to the final design stage, that the product design is integrated with the manufacturing concept, and that the manufacturing concept can proceed to the manufacturing system design phase. The review assesses whether the product and manufacturing system performance requirements can be achieved within the project budget, schedule, risk and other constraints.

Membership:

- Senior management.
- Executive champion (if appropriate).
- Project manager.
- Marketing representatives.
- Client.
- Team leaders.
- Chief designers.
- Chief co-designers.
- Designers (product).
- Critics (internal and external): External experts should be used to provide independent critique of the design concept.
- Inventors (if appropriate).
- One librarian (to record action items).

Success criteria: Success criteria include affirmative answers to the following exit questions:

- Does the current status of the technical effort and design indicate successful production of the product?
- Have the design and manufacturing configuration baselines been established and documented to enable the next phase to proceed with proper configuration management?

- Are the risks known and manageable for the next phase to proceed?
- Can the project be completed within the proposed timeline and budget?
- Are the quality and critique plans adequate for the project to proceed to the next phase?
- Has the marketing plan been established and documented in sufficient detail to proceed to the next phase?

Documentation: Record the review panel decision and get all members of the panel to sign off.

Final Design and Manufacturing System Design Phase

Goal: To generate the final product design integrated with the manufacturing system design.

Description: Finalise the physical architecture, sub-unit design, design verification, integration and design of the manufacturing system.

Inputs:

- Deliverables from the preliminary design and manufacturing concept phase.
- Source: Preliminary design and manufacturing concept review team.

Tasks and accountabilities matrix:

TABLE C.4

Tasks and Accountabilities Matrix for Final Design and Manufacturing System Design Phase

Task	Team Role Accountable for the Task
Project management	Project manager, team leaders for the product and manufacturing system design teams
Quality management	Chief designers for the product and manufacturing system design teams
Risk management	Team leader and the chief designers
Design management	Chief designers and librarians
Final design (product)	Chief designer (product), client
Manufacturing system design	Chief designer (manufacturing system), client
Critique	Critics

Useful tools:

- Modelling tools for design (e.g., CAD, physical mock-ups, etc.)

- Critiquing tools
  - Finite element analysis (FEA)
  - Failure mode effects analysis (FMEA)
  - Worst-case analysis
  - Fault tree analysis
  - Sneak analysis
  - Safety analysis
- Risk management tools
  - Contingency planning
  - De-scope planning
  - Watch lists
  - Critical items and issue list
- Modelling tools for industrial design
- Design for manufacturing and assembly (DFMA)
- Value analysis/value engineering
  - Standardisation
    - Similarity analysis across products and product families
- From-to chart
- Flow process chart
- Relationship diagram
- Timescaled Gantt chart
- Checklists and templates

Deliverables (final design):

- Finalised design configuration baseline, i.e., product structure
- Finalised bill of materials, including the coding for the ERP/MRP
- Sales configurator
- Sales brochures and other relevant marketing information
- Updated and finalised marketing plan
- Finalised industrial design
- Finalised assembly model or drawings (product structure)
- Finalised sub-unit models or drawings
- Manufacturing drawings
- Finalised product integration plan
- Finalised product documentation
- Updated risk plan
- Updated quality plan

- Design critique documentation
- All open design issues closed

Deliverables (manufacturing system design):

- Environmental plan
- Selection or design of the equipment (excluding tools, etc.)
- Factory and cell layouts (manufacturing system design configuration baseline)
- Detailed throughput time analysis
  - Time analysis
  - Throughput analysis (production rate)
- Detailed maintenance concept
- Detailed manufacturing quality plan
- Production scheduling and control system design
  - Storage and buffer planning
  - Item-level definition of material control principles
- Labour estimates
- Preliminary production ramp-up requirements
- Updated manufacturing capital estimate
- Preferred supplier list

Final Design Review (FDR)

Description: The review assesses the completeness and integration of the product and manufacturing system designs to determine whether to proceed to pilot plant trials.

Membership:

- Executive champion (if appropriate).
- Project manager.
- Client.
- Team leaders.
- Chief designers.
- Chief co-designers.
- Designers.
- Critics (internal and external): External experts should be used to provide independent critique of the design concept.

- Inventors (if appropriate).
- One librarian (to record action items).

Success criteria: Success criteria include affirmative answers to the following exit questions:

- Does the current status of the technical effort and design indicate successful production of the product?
- Have the design and manufacturing configuration baselines been updated and documented to enable the next phase to proceed with proper configuration management?
- Are the risks known and manageable for the next phase to proceed?
- Can the project be completed within the proposed timeline and budget?
- Are the quality and verification plans adequate for the project to proceed to the next phase?
- Have the key product characteristics having the most impact on system performance, assembly, cost, reliability and safety been identified?
- Have the critical manufacturing processes that impact the key product characteristics been identified and their capability to meet design tolerances determined?
- Are the sales brochures and marketing information available and suitable to allow marketing to begin?
- Has the marketing plan been updated and finalised?

Documentation: Record the review panel decision and get all members of the panel to sign off.

Pilot Trial and Operations Planning Phase

Goal: To verify in-plant production on prototype equipment and prepare operations planning documents.

Description: Build and evaluate pilot plant and product prototypes, prepare operations planning documents, carry out a producibility analysis, and design of production tools, jigs, fixtures and the work environment.

Inputs:

- Deliverables from the final design and manufacturing system design phase.
- Source: Final design and manufacturing system review panel.

Tasks and accountabilities matrix:

TABLE C.5

Tasks and Accountabilities Matrix for Pilot Trial and Operations Planning Phase

Task	Team Role Accountable for the Task
Project management	Project manager, team leader (manufacturing system)
Quality management	Chief designer (manufacturing system)
Risk management	Team leader and the chief designer for manufacturing system
Design management	Chief designer and librarian for manufacturing system
Detailed design of manufacturing cells and design and manufacture of tools, fixtures etc.	Chief designer (manufacturing system), client
Develop ramp-up schedule and training material	Chief designer (manufacturing system)
Critique	Critics

Useful tools:

- Time measurement (stopwatch and work sampling)
- 7 samurais
- Timescaled Gantt chart
- Checklists and templates

Deliverables:

- Preliminary operations sheets (methods)
  - Planning of work content
  - Skill chart templates
  - Detailed manufacturing instructions
- Throughput time Gantt chart
  - Final operation sequencing and definition of station-specific work contents (= the rake)
- Updated labour estimates
- Revised ramp-up schedule
- Design and manufacture of prototype lasts and moulds
- Design and manufacture of prototype tools and tool systems
- Design and manufacture of prototype fixtures, fasteners and jigs
- Design of production lasts and moulds
- Design and standardisation of production tools and tool systems
- Design of production fixtures, fasteners and jigs
- Initial mistake proofing (poka-yoke, etc.)

- Working environment design (5S)
- Preliminary occupational safety design
- Test reports
  - Internal
  - External
- Training material

Critical Producibility Review (CPR)

Description: The review assesses the producibility of the product, the capability of the manufacturing system to make the product within the design specification and the readiness to proceed to building and installing final production equipment.

Membership:

- Executive champion (if appropriate).
- Project manager.
- Client.
- Team leader (manufacturing system).
- Chief designer (manufacturing system).
- Chief co-designer (manufacturing system).
- Designers (manufacturing system).
- Critics (internal and external): External experts should be used to provide independent critique of the design concept.
- Inventors (if appropriate).
- Librarian (to record action items).

Success criteria: Success criteria include affirmative answers to the following exit questions:

- Have the key product characteristics affecting producibility been identified?
- Do the pilot plant trials indicate the manufacturing system has the capability to meet product specifications at full-scale production rates?
- Are mistake proofing and safety analyses sufficiently detailed to permit reliable and safe production?
- Are the operations planning documents sufficiently detailed to enable the next phase to proceed?

- Have the design and manufacturing production configuration baselines been established and documented to enable the next phase to proceed?
- Are the risks known and manageable for the next phase to proceed?
- Can the project be completed within the proposed timeline and budget?
- Are the quality and critique plans adequate for the project to proceed to the next phase?
- Have the production ramp-up requirements and schedule been established?

Documentation: Record the review panel decision and get all members of the panel to sign off.

Build and Install and Ramp-Up Phase

Goal: To finalise the manufacturing system design and operation, and ramp-up to full production rates.

Description: Build and install production equipment, carry out production trials, ramp-up to full production rates and prepare the manufacturing system readiness report.

Inputs:

- Deliverables from the pilot plant trials and operations planning phase.
- Source: Pilot plant trials and operations planning review panel.

Tasks and accountabilities matrix:

TABLE C.6

Tasks and Accountabilities Matrix for Build and Install and Production Trials Phase

Task	Team Role Accountable for the Task
Project management	Project manager, team leader (manufacturing system)
Quality management	Chief designer (manufacturing system)
Risk management	Team leader and the chief designer for manufacturing system
Updated product design	Chief designer (product)
Updated manufacturing system design, preparation of final operations planning documents	Chief designer (manufacturing system), librarian (manufacturing system)

(Continued)

TABLE C.6 (*Continued*)

Tasks and Accountabilities Matrix for Build and Install and Production Trials Phase

Task	Team Role Accountable for the Task
Construction and commissioning of a fully operational manufacturing system and preparation of manufacturing system readiness report	Project manager, team leader (manufacturing system), chief designer (manufacturing system), client, librarian (manufacturing system)
Critique	Critics

Useful tools:

- Time measurement (stopwatch and work sampling)
- 7 samurais
- Timescaled Gantt chart
- Checklists and templates

Deliverables:

- Updated product design (product production configuration baseline)
- Updated manufacturing system design (manufacturing system production configuration baseline)
- Fully operational manufacturing system
- Finalised operations planning documents
- Manufacturing system readiness report

Operational Readiness Review (ORR)

Description: The review assesses the readiness to make products at full production volume and production cycle target times.

Membership:

- Senior management.
- Executive champion (if appropriate).
- Project manager.
- Client.
- Team leader (manufacturing system).
- Chief designer (manufacturing system).
- Chief co-designer (manufacturing system).
- Chief designer (product).

- Designers (manufacturing system).
- Critics (internal and external): External experts should be used to provide independent critique of the design concept.
- Librarian (to record action items).

Success criteria: Success criteria include affirmative answers to the following exit questions:

- Do the production trials indicate the capability of the manufacturing system to produce products within specification?
- Are the staff sufficiently trained in the new production procedures?
- Do the production trials indicate the capability of the manufacturing system to produce at full production rates in a reliable and safe manner?
- Have all development issues been closed?
- Has all documentation been updated and finalised?

Documentation: Record the review panel decision and get all members of the panel to sign off.

Appendix D: Throughput Lead Time Analysis

Throughput lead time is a measure of how long it takes to make one batch of products to be completed under normal factory operating conditions. Lead time depends on the manufacturing product structure, the batch size and the normal time to make one batch of the product.

Throughput lead time analysis is best carried out early in the design process while there is still an opportunity to change the design at little extra cost. It is aimed at determining whether the product architecture is suitable to meet the delivery lead times expected by customers.

Lead time analysis will be illustrated by using a wooden table. The table construction is depicted in Figure D.1. The manufacturing product structure is shown in Figure D.2. The number of parts and supply volumes are shown in brackets.

The wood for the legs and rails is outsourced to a supplier. The wood is delivered in 10 m lengths and the lead time is 2 weeks. The wood for the tops is also outsourced to another supplier. The blanks for the tops are delivered as slabs and the lead time is 21 working days (4 weeks plus 1 day).

The machining of the legs, rails and tops takes 2 days. Leg assembly takes 1 day, and final assembly with polishing takes 2 days. The lead times can be depicted on a lead time Gantt chart (Figure D.3).

So far we have only considered manufacture and assembly. However, the customer lead time includes order gathering and delivery. For batch manufacturing, customer orders are not filled immediately unless a deliver-to-order cycle is being used. For other cycles, the customer orders are collated over a period in order to reach the minimum batch size number (order gathering). In our example, the time period is 1 week. The delivery lead time is 1 week.

We are now in a position to determine the effects of different customer lead time expectations. First, if customers are happy to accept a 7-week lead time, then the tables can be made-to-order and there is no need to carry any final or in-process inventory (assuming no rejects during manufacturing). This situation is shown in Figure D.4; the representation of the lead times is indicated in Table D.1.

The longest lead time branch is for the table tops, and this is the critical path, with a total lead time of 7 weeks.

Now consider what happens if customers demand a 5-week delivery time. The cheapest solution to meet this deadline is shown in Figure D.5. The slabs for the table are ordered separately and held in inventory to meet demand. This removes the slabs from the order cycle.

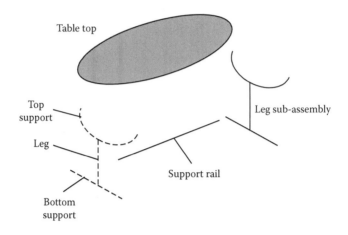

FIGURE D.1
Table construction.

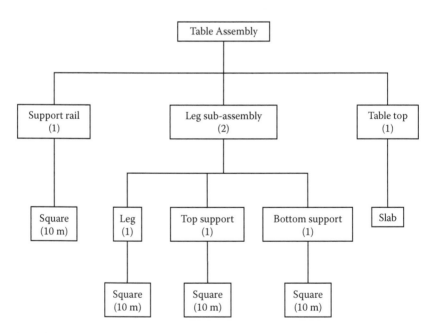

FIGURE D.2
Manufacturing product structure.

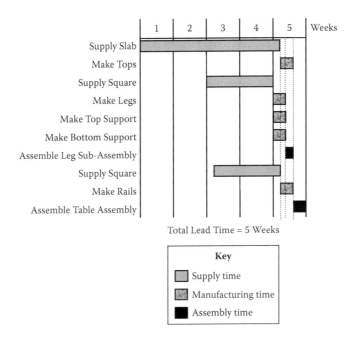

FIGURE D.3
Lead time Gantt chart.

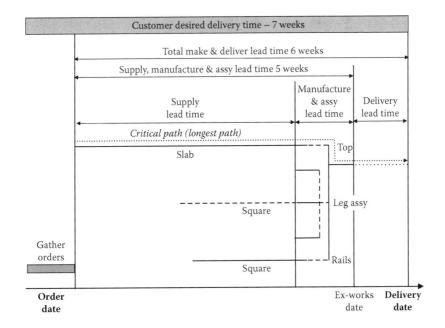

FIGURE D.4
Order cycle for 7-week delivery expectation.

TABLE D.1

Lead Time Representation for Figure D.4

Activity	Representation (line style)
Delivery	Dotted
Final assembly	Solid
Leg assembly	Dashed
Top manufacture	Dashed
Rail manufacture	Dashed
Leg manufacture	Solid
Slab supply	Solid
Square supply (legs)	Dashed
Square supply (rails)	Solid

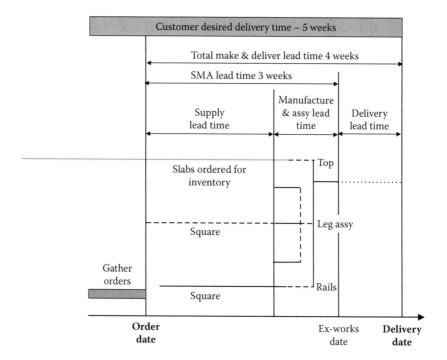

FIGURE D.5

Order cycle for 5-week delivery expectation.

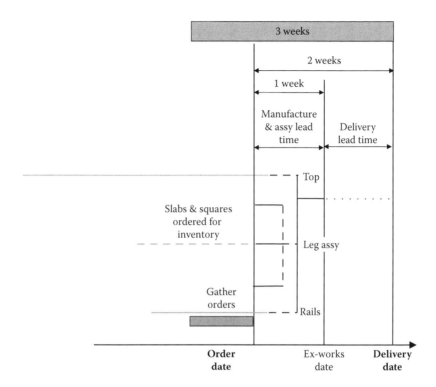

FIGURE D.6
Minimum lead time for make-to-order.

The shortest possible lead time for a make-to-order cycle is 3 weeks (Figure D.6). If customers require shorter lead times, then the remaining options are assemble-to-order and deliver-to-order, both of which result in higher inventory costs.

Other options available during the design stage to reduce lead time are:

- Re-design the manufacturing processes to reduce the batch size.
- Change suppliers to reduce supply lead time.
- Re-design the product using different materials.

Index